APPLIED GEOCHEMISTRY

AF836586

APPLIED GEOCHEMISTRY

Advances in Mineral Exploration Techniques

ATHANAS SIMON MACHEYEKI
Earth Sciences Institute of Shinyanga, Shinyanga, United Republic of Tanzania

XIAOHUI LI
School of Resources and Environmental Engineering, Hefei University of Technology, Hefei, P.R. China

DALALY PETER KAFUMU
Earth Sciences Institute of Shinyanga, Shinyanga, United Republic of Tanzania

FENG YUAN
School of Resources and Environmental Engineering, Hefei University of Technology, Hefei, P.R. China

ELSEVIER

Elsevier
Radarweg 29, PO Box 211, 1000 AE Amsterdam, Netherlands
The Boulevard, Langford Lane, Kidlington, Oxford OX5 1GB, United Kingdom
50 Hampshire Street, 5th Floor, Cambridge, MA 02139, United States

Copyright © 2020 Elsevier Inc. All rights reserved.

No part of this publication may be reproduced or transmitted in any form or by any means, electronic or mechanical, including photocopying, recording, or any information storage and retrieval system, without permission in writing from the publisher. Details on how to seek permission, further information about the Publisher's permissions policies and our arrangements with organizations such as the Copyright Clearance Center and the Copyright Licensing Agency, can be found at our website: www.elsevier.com/permissions.

This book and the individual contributions contained in it are protected under copyright by the Publisher (other than as may be noted herein).

Notices
Knowledge and best practice in this field are constantly changing. As new research and experience broaden our understanding, changes in research methods, professional practices, or medical treatment may become necessary.

Practitioners and researchers must always rely on their own experience and knowledge in evaluating and using any information, methods, compounds, or experiments described herein. In using such information or methods they should be mindful of their own safety and the safety of others, including parties for whom they have a professional responsibility.

To the fullest extent of the law, neither the Publisher nor the authors, contributors, or editors, assume any liability for any injury and/or damage to persons or property as a matter of products liability, negligence or otherwise, or from any use or operation of any methods, products, instructions, or ideas contained in the material herein.

British Library Cataloguing-in-Publication Data
A catalogue record for this book is available from the British Library

Library of Congress Cataloging-in-Publication Data
A catalog record for this book is available from the Library of Congress

ISBN: 978-0-12-819495-9

For Information on all Elsevier publications
visit our website at https://www.elsevier.com/books-and-journals

Publisher: Candice Janco
Acquisitions Editor: Amy Shapiro
Editorial Project Manager: Megan Ashdown
Production Project Manager: Vijayaraj Purushothaman
Cover Designer: Dalaly P. Kafumu and Victoria Pearson

Typeset by MPS Limited, Chennai, India

Working together to grow libraries in developing countries
www.elsevier.com • www.bookaid.org

Dedication

To all mineral exploration scientists

Contents

Foreword

Successful exploration for mineral resources depends on integrated application of our understanding of the geology that hosts ore deposits, and their geochemical and geophysical expressions. Geochemistry is one of the oldest exploration techniques, but it is also one of the most modern. In its earliest form, it involved tracing alluvial minerals, especially gold, back to their source. Over time, it has developed as a result of improved analytical techniques, different sampling media, changes in extraction methods, and improvements in data processing and presentation to become the versatile and sophisticated suite of methods available to modern explorers.

For many explorers, it is a challenge to keep up with developments in geological understanding and developments in exploration technology that can now be applied in the search for deposits. New techniques open opportunities to discover deposits that were not recognizable with previous generations of knowledge and exploration tools. Explorers who take advantage of new developments have the opportunity to be ahead of their rivals in the highly competitive business of mineral exploration.

This book takes a holistic approach to applying geochemical methods: first understand the deposits you want to find—their host geology, the characteristics of the ores, and what surface processes do to them. Then understand the behavior of chemical elements and how their distribution can be used in exploration, applying both conventional and unconventional methods across the full spectrum of available tools.

This book is for all explorers: newcomers wanting to learn the basic principles and traditional methods, and experienced explorers wanting to learn about new and innovative methods. Some developments in geochemistry improve the effectiveness of traditional techniques that have been used for decades. Others are unconventional, applying new approaches to the challenges of discovery. Awareness of the diversity of tools that are available for explorers is essential if they are to achieve the goal of ore deposit discovery by the fastest, cheapest, and most effective methods available.

Noel White

Preface

In the 1990s, the first author was field-attached to the Kabanga Nickel project as an undergraduate geology student from the University of Dar Es Salaam. The project, having two deposits by then (Kabanga Main and Kabanga North) is located about 200 km SW of Lake Victoria in Tanzania. He was surprised to learn that although most of the drill holes from the Kabanga Main deposit intersected mineralization at few meters to few hundred meters, one drill hole went as far as 1.2 km without intersecting mineralization. The geologist who cited the borehole was troubled as he could not understand why! The first author then wanted to know what exactly caused the wrong citing of the drill hole and the curiosity to solve this question grew in time.

A few years later, he was appointed by his then new employer, Anglo American Exploration Company to attend the 11th Geological Society of Africa Conference in Cape Town where, he visited various geological sites including the Volcanogenic Massive Sulfides (VMS) deposits and the Komatiites within the Kaapvaal Craton, South Africa. Around the same time, he had visited the Great Dyke of Zimbabwe and developed interest on magmatic deposits and how to explore the associated minerals—thanks to Murray Surtees the former East and Central Africa Director for Anglo America for supporting the Master of Science (MSc) studies of the first author.

In the MSc studies, it was found that, geochemical data played a great role in deciphering areas of mineralization from those which are not mineralized and contributed significantly in the degree of correctness to cite a drill hole much more than the other techniques available at that time. At the end of the studies, he proposed the NW part of the Kabanga Nickel project to be of interest and so did the data in Macheyeki (2011) indicate. The results of the studies later contributed to the discovery of the Kabanga NW Nickel deposit.

Both the Kabanga North and Kabanga NW Ni–Cu deposits do have their ore bodies located much deeper (up to more than 400 m) and inclined to the north such that conventional soil sampling is affected by two main factors: firstly, the highly Nickel-leached gossans and the laterites covers that are highly enriched with Nickel. Furthermore, there is no contrast in geochemical elements obtained on top of the Ni–Cu deposit and those from barren areas. The second factor is that all elements are at very high values near and within exposed mafic–ultramafic rocks particularly in the Kabanga Main, something that does not give a real indication as to where exactly interesting areas for drilling around the Kabanga Main deposit could be.

The author's results came up with proposed lithogeochemical vectors/ratios (nonconventional exploration techniques) that could be useful in locating Ni–Cu ores within geological settings similar to that of the Kabanga Ni–Cu deposits. Few years later, the author decided to write a book on nonconventional exploration techniques to include the developed lithogeochemical vectors. He then invited interested authors—thanks to Dr. Xiaohui Li, Dr. Dalaly P. Kafumu, and Prof. Feng Yuan who joined the team and included several other nonconventional techniques that the reader is invited to read.

Case studies on 3D Mineral Prospectivity modeling in the Yangtze Metallogenic Belt and Multifractal modeling on Cu and Au mineralization in the NW Junggar area of northern Xinjiang Province by Dr. Xiaohui Li, Prof. Feng Yuan et al. provide yet important demonstration on application of nonconventional mineral exploration techniques.

Dear reader, this book starts with an introductory part where elements of geochemistry/geochemical exploration are briefly explained. The second chapter surfaces on classification of ore deposits and geochemical techniques that can be used to tell about the geochemical characteristics of some of the ore deposits. The third chapter is devoted on the principles of nonconventional exploration techniques; a little mention on conventional exploration techniques is given at the beginning of the chapter. Finally, on Chapter 4, Application of nonconventional mineral exploration techniques: case studies, are detailed. It is the belief of the authors that you will enjoy reading this book and make use of the knowledge contained herein.

Athanas S. Macheyeki, Xiaohui Li, Dalaly P. Kafumu and Feng Yuan

Merci!

Acknowledgments

This book is another deposit of knowledge in Applied Geochemistry to be useful to exploration geologists, university geology students, and other geoscientists. In writing this book, we have been fortunate to get help and inspiration from a lot of friends and colleagues as mentioned hereunder.

We are grateful to Professor Sospeter Muhongo and Professor Marc De Batist who encouraged and commended the idea of writing this book far back in 2010 and 2012, respectively. Special thanks goes to Sebastian Ndile from Barrick Gold (Tanzania); Mohamed Zengo; John Kalimenze and Hamis Saidi, all from the Geological Survey of Tanzania (GST); Dr. Charles Kasanzu from the University of Dar Es Salaam (UDSM); Erik Werner from RSC Mining and Mineral Exploration Company; and Mr. Fredrick Mangasini from Mineral Resources Institute who all commented on the earlier draft of the book.

We are also thankful to Hafsa Maulid from GST, Hassan Mdala from the Malawi Geological Survey, and Dr. Alisante Mshiu from UDSM who provided useful literature and guidance on the principles of remote sensing. We must also thank Selemani Silingi, a former Tutor at Earth Sciences Institute of Shinyanga and a Postgraduate student at China University of Geosciences who linked authors to work together. We are also indebted to Jean De Dieu from Rwanda, Desdery Mujuni and Jonathan Mgonja from GST, Innocent Mgeta of Shinyanga, and staff of the Mining Commission in Tanzania; and many other scientists and friends who supported us in various ways.

Honourable Dotto Biteko, Prof. Idris Kikula, Prof. Abdulkarim Mruma, Prof. Shukrani Manya and Dr. AbdulRahaman Mwanga provided moral support during this work. Paulina Awasi, Mariamagdalena Mchikirwa, Mathias Macheyeki, and his sisters Juliana and Faith cherished the plan, gave courage, participated in proofreading the manuscript, and offered moral support that has kept us enthusiastic at all time during preparation, compilation, and publication of the same. We express our appreciation to all people who, in different ways supported this endeavor.

Xiaohui and Feng were financially supported by funds from the National Natural Science Foundation of China (Grant Nos. 41820104007, 41702353, 41672069,41872247), the National Key R&D Program of China (Grant Nos. 2016YFC0600209, 2016YFC0600206), the Public Geological

Survey Program of Anhui Province (Grant No. 2009-13), and the Oversea Master Program (Grant No. MS2015HFGY026). They were also financed by the National Natural Science Foundation of China (Grant No. 41672069), the Chinese National Science and Technology Programme during the 12th Five-Year Plan Period (Grant No. 2011BAB06B01), the Program for New Century Excellent Talents in University (Grant No. NCET-10-0324), and the Featured Program for University (Grant No. TS2017HFGY029).

Thanks to Professor Dr. Swapan Kumar Haldar and anonymous reviewers for their constructive comments during the book proposal review. We are also indebted to Elsevier team for their continued technical support throughout the process of this endeavor.

List of authors and their chapter contributions

1. Chapter 1, Elements of exploration geochemistry—Dr. Athanas S. Macheyeki and Dr. Dalaly P. Kafumu;
2. Chapter 2, Types of ore deposits and their origin—Dr. Athanas S. Macheyeki and Dr. Dalaly P. Kafumu;
3. Chapter 3, Conventional and nonconventional exploration techniques–principles—Dr. Athanas S. Macheyeki;
4. Chapter 4, Application of nonconventional mineral exploration techniques: case studies—Dr. Athanas S. Macheyeki, Dr. Xiaohui Li, and Prof. Feng Yuan; and
5. Cover page design by Dr. Dalaly P. Kafumu and Victoria Pearson.

CHAPTER

1

Elements of exploration geochemistry

1.1 Introduction

Geochemistry is a branch of science that deals with the content and distribution of chemical elements in minerals, ores, rocks, soils, waters, and the atmosphere. It also studies the circulation of these chemical elements in nature based on the properties of their atoms and ions (Goldschmidt, 1954). Exploration geochemistry is a branch of applied geochemistry that uses chemical elements, properties, and distribution in nature to locate and characterize economic mineral deposit(s). In order to conceptualize the distribution and circulation of chemical elements in nature, it is practical to have a framework of the geochemical cycle between the different reservoirs and earth compartments (e.g., Lasaga and Ohmoto, 2002; Eriksson et al., 2004; Condie, 2011; Brimblecombe, 2014; Petsch, 2014).

Geochemical cycle refers to the earth's internal and external processes involving carbon, phosphorous, sulfur, water, and rock elements exchange (Fig. 1.1A and B). These processes within the earth's interior (lithosphere, crust, mantle, and core), surface and near surface (hydrosphere, biosphere, and soils), the atmosphere (troposphere, stratosphere, mesosphere, and thermosphere), and outer space (meteorites and asteroids) greatly affect the mineral formations in the geochemical cycle. Thus the study of geochemical cycle is the study of rock chemistry, soil chemistry, water chemistry, organic chemistry (or biogeochemistry), and gas chemistry. With the geochemical cycle in mind, we can be able to appreciate geochemical processes that lead to formation of mineral deposits and the methods used to explore the minerals (exploration geochemistry). In the next subchapters, emphasis is given to the use of rock and soil chemistry in mineral exploration.

Applied Geochemistry
DOI: https://doi.org/10.1016/B978-0-12-819495-9.00001-3

© 2020 Elsevier Inc. All rights reserved.

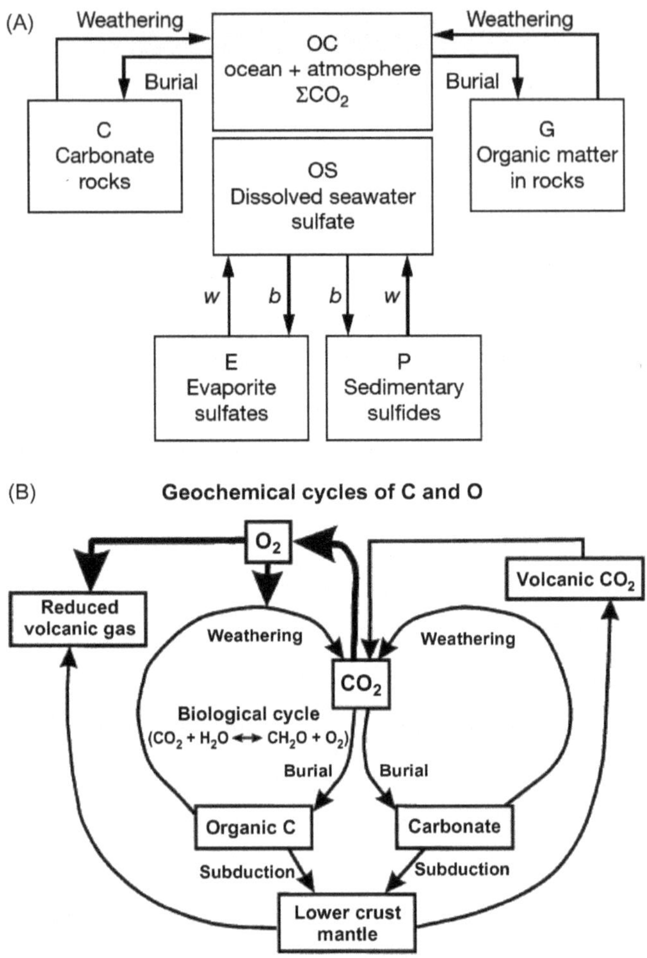

FIGURE 1.1 (A) The simplified geochemical cycles of carbon and sulfur, including burial and weathering of sedimentary carbonates, organic matter (OM), evaporites, and sulfides. The relative fluxes of burial and weathering of OM and sulfide minerals play a strong role in controlling the concentration of atmospheric O_2. (B) Schematic illustration of the geochemical cycles for carbon and oxygen. Note that the long-term production rate of atmospheric O_2 equals the burial rate of organic C in sediments, and that the atmospheric O_2 is consumed by reduced volcanic gas and by rock weathering (Lasaga and Ohmoto, 2002). *b*, Burial, *w*, weathering. Source: *(A) From Petsch, S.T., 2014. The global oxygen cycle. Treatise Geochem., 437–473. Available from: https://doi.org/10.1016/b978-0-08-095975-7.00811-1.*

1.1.1 Rocks and their types

Rocks are made up of minerals and lithified sediments. There are three main types of rocks; igneous rocks, sedimentary rocks, and

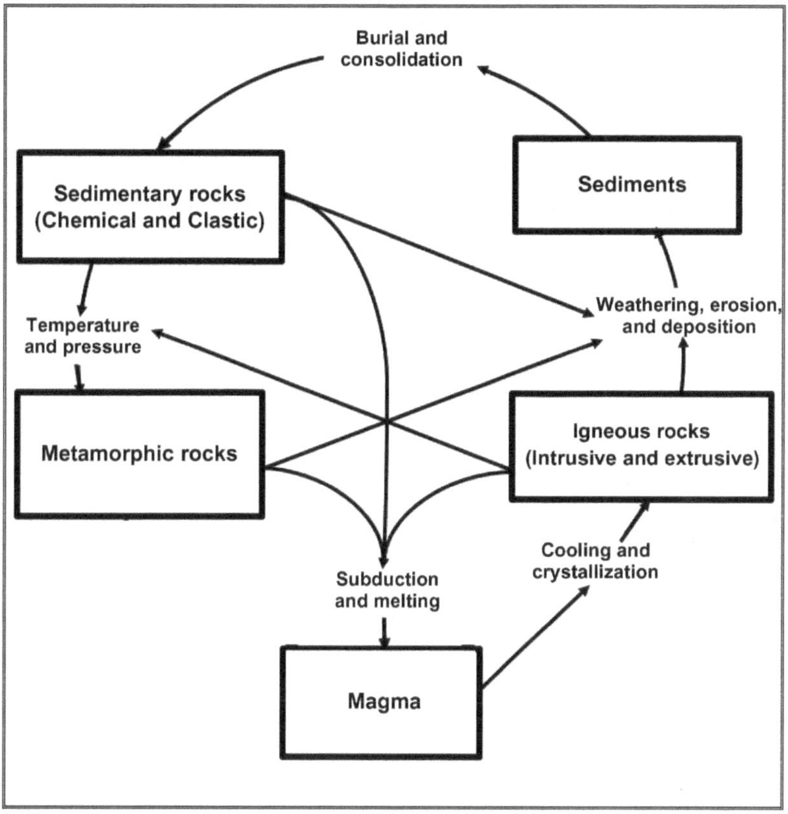

FIGURE 1.2 Idealized diagram of the rock cycle.

metamorphic rocks. In a simplified model of formation, they can be represented by the rock cycle diagram (Fig. 1.2).

The rock cycle is an idealized diagram that explains how rocks are formed and how they are related to each other in terms of processes that lead to their formations. The rock cycle genesis and origin include the formation and the relationships of igneous rocks, sedimentary rocks, and metamorphic rocks.

1.1.1.1 Igneous rocks

Igneous rocks can be classified into intrusive (plutonic) and extrusive (volcanic) rocks. Igneous rocks sometimes called magmatic rocks are formed through the cooling and solidification of magma or lava. The magma can be derived from partial melts of existing rocks in either the earth's mantle or crust. Solidification into rock occurs either below the surface (intrusive) rocks or on the surface (extrusive) rocks. Those in

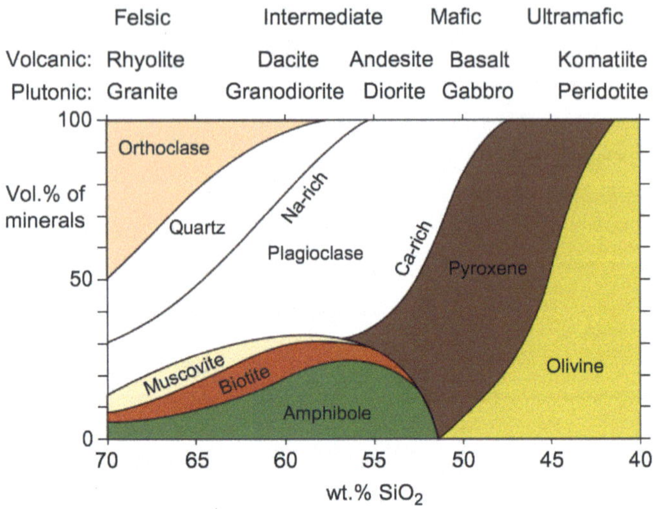

FIGURE 1.3 Mineralogical classification of common magmatic rock types. Source: *From Schön, J.H., 2015. Rocks—their classification and general properties. Physical Properties of Rocks. Fundamentals and Principles of Petrophysics, pp. 1–19.*

between are termed as intermediate (hypabyssal) rocks. Igneous (intrusive and extrusive) rocks are generally classified using grain size and mineral composition as shown in simple schematic classification that identify the main forms of igneous rocks by the variation of grain size (volume of mineral in percent) on the ordinate and mineral composition in ether increasing silica (acidic) of ferromagnesian minerals (basic) on the abscissa (Fig. 1.3).

1.1.1.1.1 Intrusive rocks

An intrusive rock is a rock whose source is "magma" and is referred to as plutonic rock. The magma must have solidified deep into the crust resulting in relatively large minerals (crystals). Classification diagram for intrusive (plutonic) rocks is shown in Fig. 1.4. This classification scheme is based on mineral content (vol.%): Q = quartz, A = alkali feldspar (including albite); P = plagioclase; F = feldspathoids; M = mafic and related minerals. Rocks with M less than 90 are named according to their positions in the QAPF diagram, the light-colored constituents being calculated to the sum 100. The following are treated: granitoids and related rocks, ultramafic and gabbroic rocks, charnockitic rocks, feldspathoidal rocks (Streckeisen, 1976).

Further references on igneous rocks include Cox et al. (1979), Cas and Wright (1987), Le Bas and Streckeisen (1991), Rollinson (1993), Bellieni et al. (1995), Gillespie and Styles (1999), and Schön (2015).

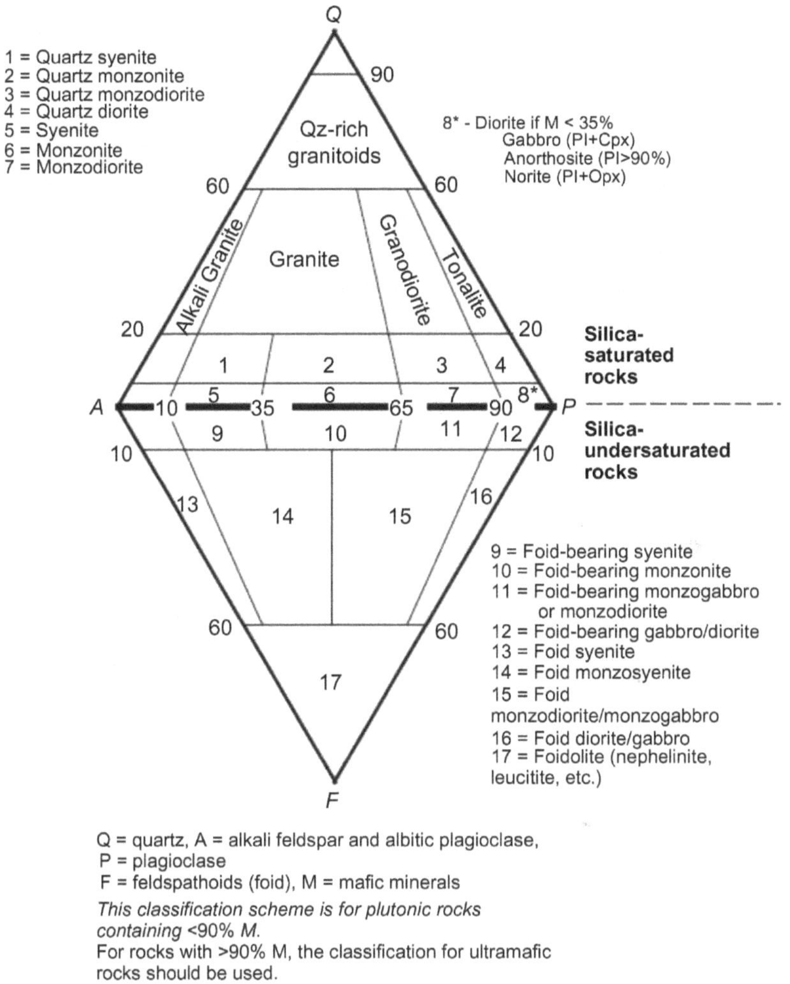

FIGURE 1.4 General classification and nomenclature of plutonic rocks according to mineral content (in vol.%). Q + A + P = 100, or A + P + F = 100 (Streckeisen, 1976).

The basic to ultrabasic or mafic to ultramafic igneous rocks are rich in dark-colored alkali minerals like feldspathoids, aegirine, and sodic amphiboles all enriched in alkalis elements like iron, sodium, and potassium. Common mafic (basic) rocks include gabbro and dolerite. Acidic (felsic) rocks are relatively rich in silicate minerals like quartz, muscovite, orthoclase, and the sodium-rich plagioclase feldspars (albite-rich). They are enriched with lighter elements such as silicon, oxygen, and aluminum. Common felsic intrusive rocks include granites, diorites, and granodiorites.

1.1.1.1.2 Extrusive rocks

An extrusive (volcanic) rock as is for the intrusive rock its source is "magma" and the magma must be forced to the surface in volcanic eruptions and then solidifies. The texture of an extrusive igneous rock is fine due to a rapid solidification of magma resulting in relatively fine mineral matrix. Extrusive rocks, like intrusive rocks can be classified based on $Na_2O + K_2O$ (wt.%) versus SiO_2 (wt.%), Fig. 1.5. As for intrusive rocks, there is no single classification scheme that can be singled out to fully suite extrusive rocks classification scheme. Fig. 1.5 provides the basic classification scheme thus Le Bas and Streckeisen (1991), Rollinson (1993), Bellieni et al. (1995), Gillespie and Styles (1999), and Schön (2015) are examples of publications that can provide more details.

1.1.1.1.3 Hypabyssal rocks

Therefore hypabyssal rocks are igneous intrusive rocks that crystallized at intermediate depths resulting in medium grain sizes and textures. They are subvolcanic rocks formed near surface where temperatures are relatively low and the cooling of the magma is neither rapid nor slow (Gillespie and Styles, 1999). Examples of hypabyssal rocks include microgranites, porphyry, microgranodiorites, microdiorites, and dolerites.

1.1.1.2 Sedimentary rocks

Sedimentary rocks are formed due to weathering and decomposition of any type of rock (igneous, sedimentary, or metamorphic) on the earth's surface forming sediments. The sediments can be deposited in situ, transported, or redeposited by ice, wind, gravity, or water flows carrying sand, gravels, and boulders or in suspension or chemical solutions. Sedimentary rocks start with weathered sediments that are then deposited and undergo burial, lithification (hardening), and diagenesis is under favorable conditions to form sedimentary rocks (Kisch, 1979). In other words, weathered sediments undergo burial, diagenesis, consolidation, and cementation inducing physical and chemical changes that convert the sediments into sedimentary rocks of the earth crust.

Sedimentary rocks are classified in four main types based on the processes responsible for their formation as: clastic sedimentary rocks; biochemical (biogenic) sedimentary rocks; chemical sedimentary rocks (Pettijohn, 1983); and volcanogenic sedimentary rocks formed by impacts, volcanism, and other minor processes. The classification of sedimentary rocks is discussed in the proceeding subtopics.

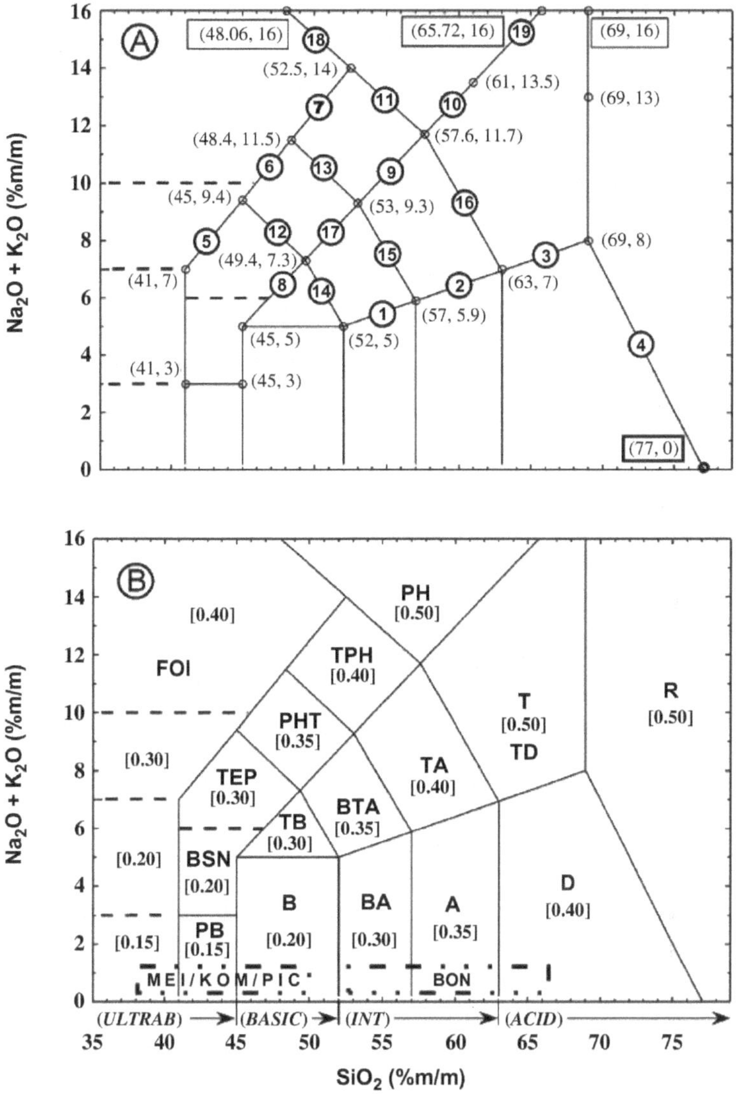

FIGURE 1.5 TAS classification scheme ($Na_2O + K_2O$ vs SiO_2, both parameters in %m/m unit, a correct term now recommended to be used for wt.%) for volcanic rocks recommended by Le Bas et al. (1986), Le Bas (1989), and Le Bas (2000). (A) Note modifications of *X*- and *Y*-axis limit from original TAS diagram used in SINCLAS (coordinates in boxes). Boundary lines and their coordinates shown in parentheses (*x*, *y*) are used for programming of classification procedures. Numbers in boldface and in circles refer to equations listed in Table 1.2. (B) Field names are as follows: *A*, andesite; *B*, basalt; *BA*, basaltic andesite; *BSN*, basanite; *BTA*, basaltic trachyandesite; *D*, dacite; *FOI*, foidite; *PB*, picrobasalt; *PH*, phonolite; *PHT*, phonotephrite; *R*, rhyolite; *T*, rrachyte; *TA*, rrachyandesite; *TB*, rrachybasalt; *TD*, rrachydacite; *TEP*, rephrite; *TPH*, rephriphonolite. Fractional values ([0.15]–[0.50]) in boldface–italics and in large parentheses are Fe_2O_3/FeO ratios recommended by Middlemost (1991) for different rock types. Also included are high-Mg and picritic volcanic rocks (Le Bas, 2000). *MEI*, meimechite; *KOM*, komatiite; *BON*, boninite; *PIC*, picrite. Subdivision of magma types ultrabasic (ULTRAB), basic (BASIC), intermediate (INT), and acid (ACID) is also shown (Le Bas, 2000; Verma et al., 2002).

1.1.1.2.1 Clastic sedimentary rocks

Clastic sedimentary rocks are composed of other rock fragments that were cemented by silicate or calcite minerals. Clastic rocks are composed of largely quartz, feldspar, rock (lithic) fragments, clay minerals, and mica. Any type of mineral may be present, but they represent in general the minerals that existed in the locality. Clastic sedimentary rocks are subdivided according to the dominant particle size. Grain size scale is used to divide sediment into three fractions: gravel (≥2 mm diameter), sand (0.0–2 mm diameter), and mud or clay (≤0.0625 mm) and silt (between 0.0625 and 0.0039 mm). The classification of clastic sedimentary rocks according to particle sizes are therefore: conglomerate and breccia sedimentary rocks made of gravel; sandstone sedimentary rocks made of sand; and mudstone/claystone/siltstone/shale sedimentary rocks made of the finest materials (silt and mud).

Conglomerates are mainly composed of rounded gravel, while breccias are composed of mainly angular gravel. Sandstones are mainly composed of rounded or angular quartz, feldspar, and lithic grains cemented by muddy matrix of either silica of calcite between the larger grains. Mudstones/claystones/shales are sedimentary rocks composed of at least 50% silt- and clay-sized particles. These relatively fine-grained particles are commonly transported as suspension by turbulent wind or water flow and deposited in basins. The classification of clastic sedimentary rocks is as indicated in Table 1.1 (see Table 1.2).

1.1.1.2.2 Chemical and biological sedimentary rocks

Chemical sedimentary rocks are formed when mineral constituents in solution become supersaturated and chemically precipitate. Common

TABLE 1.1 Classification of clastic sedimentary rocks based on grain size.

Types of sediments	Grain size		Type of rock
Gravels	Rounded fragments	Course ≥ 2 mm	Conglomerate
	Angular fragments	Course ≥ 2 mm	Breccia
Sand	Rounded or Angular	Medium ≥ 0.0625 ≤ 2 mm	Sandstone
Clay	Mud	Fine ≥ 0.0039 ≤ 0.0625 mm	Siltstone
	Mud	Very fine ≤ 0.0039 mm	Mudstone
			Claystone
			Shale

After Schön, J.H., 2015. Rocks—their classification and general properties. Physical Properties of Rocks. Fundamentals and Principles of Petrophysics, pp. 1–19.

TABLE 1.2 Equations used by SINCLAS to define some TAS fields during rock classification procedures.[a]

Equation number	Intercept (B)	Slope (M)
1	− 4.36000000	0.180000000000
2	− 4.55000000	0.183333333333
3	− 3.50000000	0.166666666666667
4	− 77.00000000	− 0.10000000000000
5	− 17.6000000	0.600000000000000
6	− 18.39141765	0.617647058800
7	− 18.01219512195	0.609756097561
8	− 18.52272727273	0.5227272727273
9	− 18.35217391304	0.5217391304348
10	− 18.79411765	0.529411764700
11	37.67647059	− 0.450980392200
12	30.87727272	− 0.477272727000
13	34.64782606	− 0.478260869000
14	51.00000000	− 0.884615384000
15	54.35000000	− 0.850000000000
16	61.83333334	− 0.870370370400
17	− 20.144444444445	− 0.555555555556
18	37.64864865	− 0.450450450500
19	− 80.93220318	− 0.529661016900

[a]*Equation numbers correspond to numbers in circles in Fig. 1.5. Two complementary sets of equations are used:* $y = mx + b$ *or* $x = (y - b)/m$.
From Verma, S.P., Torres-Alvarado, I.S., Sotelo-Rodríguez, Z.T., 2002. SINCLAS: standard igneous norm and volcanic rock classification system. Comput. Geosci. 28 (5), 711–715. Available from: https://doi.org/10.1016/s0098-3004(01)00087-5.

chemical sedimentary rocks include limestone, evaporites, chert, and banded iron formations (BIFs). Limestone is formed from biological and chemical processes. Almost all limestone forms in the oceans, and most of that forms on the shallow continental shelves, especially in tropical regions with coral reefs. Reefs are highly productive ecosystems populated by a wide range of organisms, many of which use calcium and bicarbonate ions in seawater to make carbonate minerals (especially calcite) for their shells and other structures. Limestones are made up almost entirely of fragments of marine (ocean) organisms that produce calcite ($CaCO_3$) minerals for their shells and other hard parts. Limestone made of calcite and dolomite can also be formed by crystallization in saline lacustrine (lakes), tufa hot spring emanations forming travertine, and cave speleothems (stalactites and stalagmites) (Dunham, 1962).

Evaporites chemical sedimentary rocks are formed in inland lakes that have no stream outlet and the water that flows into these lakes slowly evaporates and the water becomes increasingly concentrated with dissolved salts, and eventually saturates and then crystallizes into chemical sedimentary rocks. Examples of evaporates include Chlorides: halite ($NaCl$), sylvite (KCl), and carnallite ($KMgCl_3$); Sulfates: gypsum ($CaSO_4 \cdot H_2O$), anhydrite ($CaSO_4$), and kieserite ($MgSO_4$); and Carbonates: dolomite ($CaMg(CO_3)_2$), calcite ($CaCO_3$), and magnesite ($MgCaCO_3$) (Boggs, 2006; Jackson, 1997).

Chert chemical–biological sedimentary rocks are formed by biological activities of marine organisms like radiolaria and diatoms. Diatoms and radiolaria use silica (SiO_2) to build their tiny shells and when they die their shells settle slowly to the bottom where they accumulate as layered sedimentary chert. BIFs were formed in deep seafloor environments where iron rich in iron ions seawater was oxidized to form layered iron oxide (Fe_2O_3) in form of BIFs (Hoffman et al., 1998). The origin and genesis of the BIF is associated with photosynthetic bacteria known as cyanobacteria (blue-green algae) that consumed carbon dioxide from the atmosphere and used solar energy to convert it to oxygen in the Precambrian, 2.4 billion years ago. Free oxygen levels increased in the atmosphere and the oceans and gradually converted soluble ferrous iron (Fe^{2+}) to insoluble ferric iron (Fe^{3+}), which combined with oxygen to form the mineral hematite (Fe_2O_3), leading to the accumulation of BIFs (Lyons and Reinhard, 2009).

1.1.1.2.3 Volcanogenic sedimentary rocks

The fourth category of sedimentary rocks which are formed by pyroclastic volcanic flows, impact breccia, volcanic breccia, and tuffs deposited in sedimentary basins and usually occurs in alternation with other conventional sedimentary rocks like limestone, mudstone, sandstones, and others. These rocks are subdivided into volcanogenic-clastic and chemogenic consisting of volcanic and sedimentary materials occurring together and they can be solid and detrital or chemical. Detrital or pyroclastic volcanic materials can be cemented into the place where it fell or deposited. Detrital volcanogenic sedimentary rock is divided according to size into very coarse (with particles larger than 10 mm), coarse (2–10 mm), medium (0.1–2 mm), fine (0.01–0.1 mm), and very fine (smaller than 0.01 mm) (Azibekov et al., 2019). Examples of detrital volcanogenic rocks are tuffaceous sandstones and mudstone; tuffaceous gravels; tuffaceous conglomerates; tuffaceous breccia; and tuffaceous grits (Kafumu and Paepe, 2003). Chemogenic volcanogenic sedimentary rocks form as a result of precipitation of hot water carrying ions leached from volcanic rocks. These rocks precipitate within mid-oceanic ridges and rift valley systems. Examples of chemogenic volcanogenic sedimentary rocks include jaspers, phosphorites, and depositions of sulfur, iron ores, and manganese.

1.1.1.3 Metamorphic rocks

Metamorphic rocks are formed when rocks (sedimentary, igneous, and metamorphic) are subjected to high heat, high pressure, hot mineral-rich fluids or, more commonly, some combination of these factors deep within the earth or where tectonic plates meet. Metamorphism is the process of heating of the rocks at favorable temperature, pressure, and chemistry conditions without melting the rocks, but transforms them into denser, more compact rocks. During metamorphism, new minerals, textures, and structures like folds and foliations are formed either by rearrangement of mineral components or by reactions with fluids that enter the rocks.

At favorable temperature, pressure and chemistry conditions; sediments are metamorphosed into slates, phyllite/schist, to gneiss, granulite, and finally to eclogite—a process that is exemplified by pelitic rocks. Metamorphic rocks can also be classified based on their facies: zeolite—(Prehnite–pumpellyite/greenschist/blue schist)—amphibolites—gneiss—granulite—eclogite; and depending on the state of temperature and pressure, hornfels, sanidinite, and skarns can be formed (Fig. 1.6).

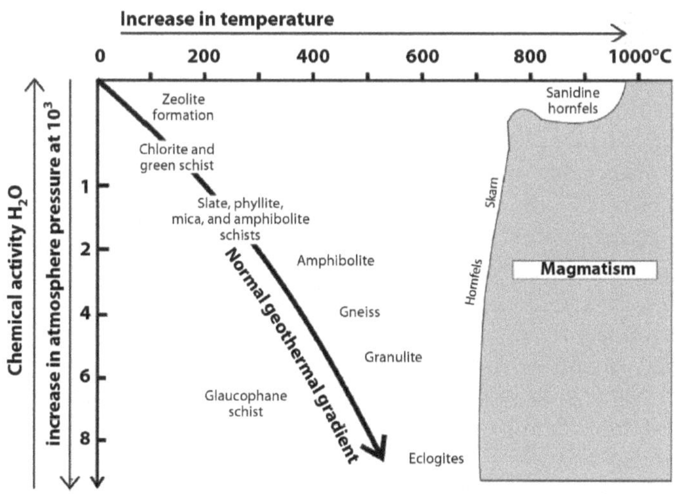

FIGURE 1.6 Metamorphic facies as a function of pressure and temperature (Haldar and Tišljar, 2013).

If metamorphic rocks are buried much deeper in the crust (Fig. 1.6), they melt and form igneous rocks. Igneous and metamorphic rocks, when exposed to erosion, form sediments that when deposited in basins followed by lithification, form sedimentary rocks again—hence the rock cycle as was indicated in Fig. 1.2.

1.1.2 Soils and regolith

Regolith is the layer of an unconsolidated rock and soil material that forms the surface of the earth and covers the bedrock. Regolith also includes soil, which is a biologically active medium and a key component in plant growth. Soil is therefore the portion of the regolith that is able to support plant and soil organism's life. Soil contains organic matter, liquids, and minerals, while most of the other layers of the regolith do not.

Regolith is a product of weathering. Bedrock may be exposed to in situ chemical weathering in which water or other compounds percolate through the rock and alter the rock's mineral content over time, breaking down some material into smaller components and separating it from the bedrock layer. Bedrock can also become regolith as a result of mechanical weathering, a process that breaks the rock into smaller pieces through the application of a force, such as thermal expansion, freeze–thaw cycles, or scouring by particles carried by wind and water. Plant roots can also assist the weathering process by penetrating and widening cracks already present in the rock.

1.1.2.1 Weathering and erosion

Weathering and erosion are two geological processes that deal with the physical and chemical changes and movement of rocks and soils on the earth's surface. Weathering is the physical or chemical breakdown of rocks, minerals, and soils due to weather conditions. Weathering occurs in situ (on site), that is, in the same place, with little or no movement. It occurs mainly through the action of wind, water, ice, chemicals, and temperature on rocks and soils. It does not involve the transport of objects to a new location.

Erosion is the removal and transport of surface materials (soil, rocks, mud, sediments, etc.) through the actions of wind, water, ice, and gravity. It can occur slowly over time, such as the migration of sand dunes, or very quickly, such as a result of flooding. Erosion involves the movement of rocks and minerals by agents of water, ice, snow, wind, waves, and gravity and transported and deposited in other locations.

There are mainly two types of weathering: mechanical or physical and chemical weathering. Mechanical or physical weathering involves the breakdown of rocks and soils through direct contact with atmospheric conditions, such as heat, water, ice, and pressure. Chemical weathering involves the direct effect of atmospheric chemicals or biologically produced chemicals also in the breakdown of rocks, soils, and minerals.

1.1.2.1.1 Physical weathering

Physical weathering, also called mechanical weathering, is a process that causes the disintegration of rocks, mineral, and soils without chemical change. The primary process in physical weathering is abrasion (the process by which clasts and other particles are reduced in size). Physical weathering can occur due to temperature, pressure, frost, root action, and burrowing animals. For example, cracks exploited by physical weathering will increase the surface area exposed to chemical action, thus amplifying the rate of disintegration.

Abrasion by water, ice, and wind processes loaded with sediment can have tremendous cutting power to forms gorges, ravines, and valleys around the world. In glacial areas, huge moving ice masses embedded with soil and rock fragments grind down rocks in their path and carry away large volumes of materials. Plant roots sometimes enter cracks in rocks and tare them apart, resulting in some disintegration and the burrowing of animals may help disintegrate rocks.

1.1.2.1.2 Chemical weathering

Chemical weathering involves the interaction of rock with mineral solutions (chemicals) to change the composition of rocks. In this process, water interacts with minerals to create various chemical reactions and transform the rocks. Chemical weathering is a gradual and ongoing process as the mineralogy of the rock adjusts to the near-surface environment. Secondary minerals develop from the original primary minerals of the rock. In this the processes oxidation and hydrolysis are the most frequent chemical processes that take place. Chemical weathering is enhanced by such geological agents as the presence of water and oxygen, as well as biological agents as the acids produced by microbial and plant root metabolism.

1.1.2.2 Soil profiles

Soil constitutes a mixture of organic matter, minerals, gases, liquids, and organisms that together support life and is a result of in situ weathering. Soil is often called the pedosphere which is the topmost layer of the earth composed of decomposed rock materials, minerals, roots, and humus (Brady and Weil, 1996; Hartemink, 2009). Soils can develop on top of bedrock (autochthonous) or can be formed elsewhere, transported and deposited on bedrock or on top of an existing soil cover in a different place (allochthonous). Soil is categorized as colluvium if it has developed at or within proximity of its provenance.

When soil has been transported by wind (air), it is called eluvium soil and when the transportation agent is water (e.g., rivers), it is referred to as alluvium. Soil develops into a profile that can vertically be subdivided into horizons, O, A, B, C, and D or R (Fig. 1.7). In the exploration of minerals, the main focus is on B-horizon since it is the horizon that is relatively richer in minerals compared to A-horizon. The C-horizon is made up of weathered/fractured materials. Below

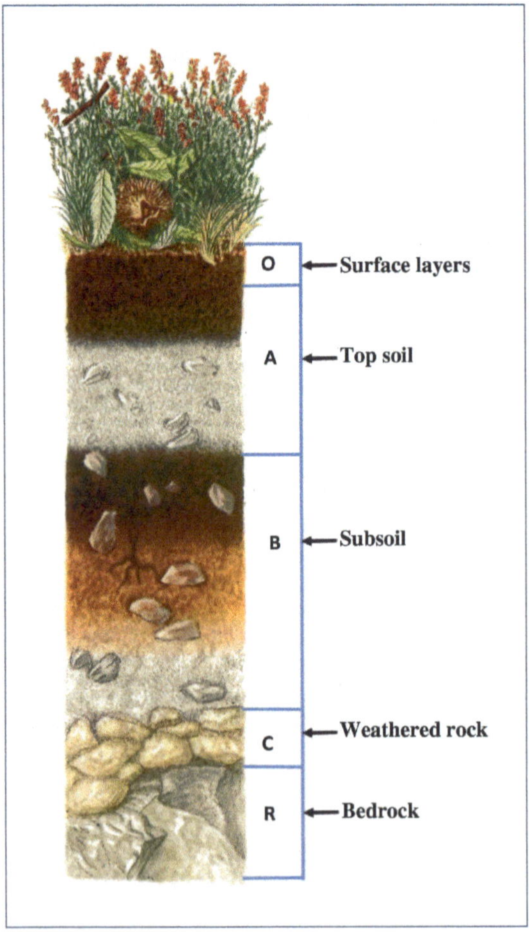

FIGURE 1.7 Soil horizons (*O*, surface layers; *A*, often rich in humus and minerals; *B*, poor in humus, rich in minerals, it is the most diverse horizon with the most subclassifications; *C*, weathered/fractured parent material; *R*, bedrock). Source: *Modified after Hartemink, A.E., 2009. The depiction of soil profiles since the late 1700s. Catena 79 (2), 113–127.*

C-horizon is the D- or R-horizon that is basically the parent rock. The near-surface materials are collectively called O-horizon. It follows therefore that soil is the product of rock weathering and its geochemical composition represents its parent rock (protolith).

1.1.3 Soil classification

Soil classification schemes are established based on soil uses and even locality/country, and so on. For instance, soil classification scheme for engineering will be different from the classification scheme for agriculture and so for mineral exploration or geology purposes (Fitzpatrick, 2004). Furthermore, one country may have its own scheme different from another country for the same uses. However, there is a Unified Soil Classification Scheme that considers soil grain sizes and texture. Regarding the classification schemes based on geology (mineral exploration) which is the scope of this book, the classification is basically a function of grain sizes; that is, course to fine (boulder to clay). One of such schemes of soil classification is that of Wentworth (1922; Table 1.3) where for instance sand grain sizes are between 2 and 0.0625 mm; silt is 0.0265–0.0039 mm; clay is everything smaller than 0.0039 mm (Fig. 1.8).

Soil classification is also represented by ternary or triangular diagram that take the percentages of sand, silt, and clay and measure them off against the tick marks. Each corner represents 100% of the grain size it is labeled with, and the opposite face of the diagram represents zero percent of that grain size (Fig. 1.9).

1.1.3.1 Regolith profiles

Regolith geochemistry is the science of studying profiles of unconsolidated (and weathered) materials that include soil cover, duricrusts, and broken rock materials. It refers to the study of the behavior of elements within a given weathering profile. It is a popular endeavor in Australia and other parts of the world that experienced deeply weathered terrain (e.g., Taylor and Eggleton, 2001). Regolith forms the interaction between weathering, erosion, transportation, and sedimentation.

From the regolith and its mode of element distribution, one can predict the trend of mineral or element concentration in that particular regolith profile. Considering Fig. 1.10, it is easy to predict elements that shall be leached, retained, or concentrated in that particular regolith profile.

Regolith profiles can be subdivided into two main parts: pedolith and saprolith (Table 1.3). Pedolith refers to the upper part of profiles composed of soil/mobile zone/biomantle (as the uppermost zone of the regolith). It may have horizons in which plant roots and organisms live

TABLE 1.3 Regolith geology profile.

Broad subdivision	Alternative terminologies				
	General term	Butt and Zeegers (1992)	Nahon and Tardy (1992)	French	CSIRO
Pedolith	Ferruginous soil (Latosol)	Soil	Soil	Soil ferrallitique	Soil
	Ferruginous zone (Laterite)	Laterite gravel	Pebbly ferruginous layer		Lateritic gravel
	(Lateritic ironstone) (Plinthite)	Cuiresse (pisolitic, nodular, massive)	Indurated conglomerated iron crust	Culrase (Ferricrete)	Lateritic duricrust (pisolitic, nodular, massive, vermiform, fragmental)
	Mottled zone	Mottled clay zone	Soft nodular crust	Carapcenodulaire	
		Plasmic/arenose horizone	Mottled (clay) zone	Argilostachetees	Mottled zone
					Plasmic/arenose horizon
					Collapsed and/or bracciated saprolite
Saprolith	Saprolite (Pallid zone)	Saprolite	Fine saprolite	Lithomarge Argilesbariolees	Ferruginous saprolite
					Clay saprolite
					Saprolite
		Saprock	Coarse saprolite	Altération pistache Arène/grus	Saprock
Protolith	Bedrock	Unweathered/fresh bedrock	Bedrock	Roche mere	Fresh rock
	Unweathered rock				

After Taylor, G., Eggleton, R.A., 2001. Regolith Geology and Geomorphology. John Wiley & Sons Ltd. 4375 pp; Macdonald, E.H., 2007. Gold deposition in the weathering environment. Handbook of Gold Exploration and Evaluation, 134–194. Available from: https://doi.org/10.1533/9781845692544.134; Pope, G.A., 2015. Regolith and weathering (rock decay) in the critical zone. In: Developments in. Earth Surface Processes, pp. 113–145. doi:10.1016/b978-0-444-63369-9.00004-5.

Millimeters (mm)		Micrometers (μm)	Phi (ϕ)	Wentworth size class	
	4096		−12.0	Boulder	Gravel
	256		−8.0	Cobble	
	64		−6.0	Pebble	
	4		−2.0	Granule	
	2.00		−1.0	Very coarse sand	Sand
	1.00		0.0	Coarse sand	
1/2	0.50	500	1.0	Medium sand	
1/4	0.25	250	2.0	Fine sand	
1/8	0.125	125	3.0	Very fine sand	
1/16	0.0625	63	4.0	Coarse silt	Silt
1/32	0.031	31	5.0	Medium silt	
1/64	0.0156	15.6	6.0	Fine silt	
1/128	0.0078	7.8	7.0	Very fine silt	
1/256	0.0039	3.9	8.0	Clay	Mud
	0.00006	0.06	14.0		

Phi (ϕ) = $-(\log_2(d))$ where d is in mm, 10^{-3} mm = 1 micron (μm)

FIGURE 1.8 Soil classification. Source: *After Wentworth, C.K., 1922. A scale of grade and class terms for clastic sediments. J. Geol. 30 (5), 377–392; Schön, J.H., 2015. Rocks—their classification and general properties. Physical Properties of Rocks. Fundamentals and Principles of Petrophysics, pp. 1–19.*

or bioturbate. Below this zone is an indurated cemented material, the duricrust. The duricrust is characterized by various fabrics and cements of Fe, Si, Ca, and Al oxides/hydroxides (Taylor and Eggleton, 2001). Just below the duricrust is the mottled zone that is composed of different colored patched materials generally red/brown within gray/white matrix.

The mottled zone is the lowest zone of the pedolith. Before going into the saprolith, there is an interface zone or boundary characterized by alternating redox conditions. Saprolith, like the pedolith, is composed of three main parts namely saprolite, saprock, and fresh rock. Saprolite is a very highly weathered to moderately weathered rock, easily broken, but retains rock fabric. Saprock is a slightly weathered rock which cannot be broken in the hand and retains rock fabric. On the other hand, the fresh rock horizon shows no signs of weathering. Other details of the parts of pedolith and saprolith are given in Table 1.3.

There are several regolith types that can be recognized in geology; the most famous ones are the laterite regoliths (Robb, 2005). Other types of regoliths are those of kaolin (Fang et al., 2019), typical of surficial weathering in a subtropical climate (Fig. 1.10).

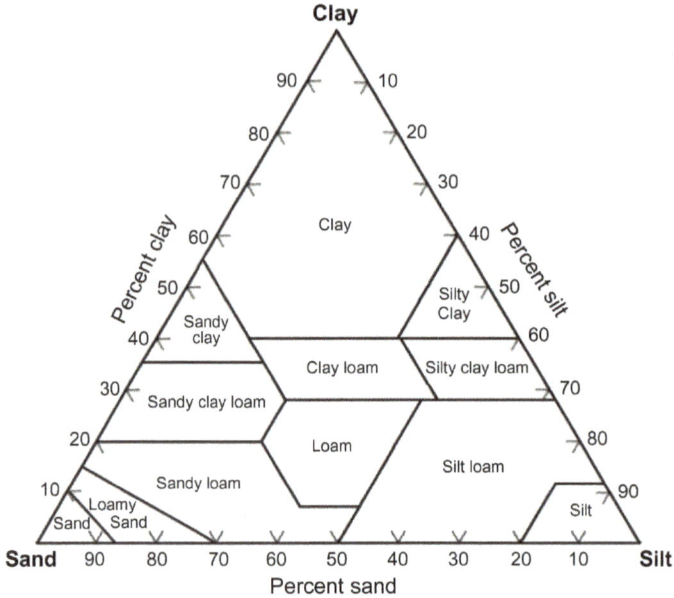

FIGURE 1.9 Ternary diagram for the classification of soils. Source: *After Shepard, F.P., 1954. Nomenclature based on sand-silt-clay ratios. SEPM J. Sediment. Res. 24, 151–158; Coleman, D.C., Callaham, M.A., Crossley, D.A., 2018. Introduction to soil. Fundamentals of Soil Ecology, pp. 1–20. Available from: https://doi.org/10.1016/b978-0-12-805251-8.00001-6.*

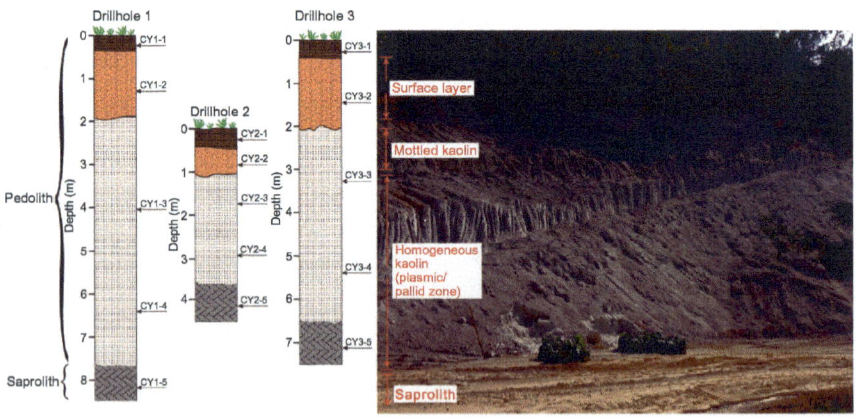

FIGURE 1.10 A generalized kaolin regolith profile showing the different horizons and the terminologies used in their description. *Source: After Fang, Q., Hong, H., Furnes, H., Chorover, J., Luo, Q., Zhao, L., et al., 2019. Surficial weathering of kaolin regolith in a subtropical climate: implications for supergene pedogenesis and bedrock argillization. Geoderma 337, 225–237. Available from: https://doi.org/10.1016/j.geoderma.2018.09.020. Compare with Table 1.3.*

1.1.3.2 *Sediments*

Sediments are loose organic or inorganic materials from any source. It is the rivers, winds, and other erosion agents that transport sediments from one place to another and deposit them in basins. Materials that are transported or deposited by river processes are called "stream sediments" and those by wind, "aeolian sediments." In mineral exploration, stream sediments are of interest because they represent a relatively large catchment area of those sediments.

Therefore it is important to know where to sample stream sediments and the reason for sampling at that locality(ies). By sampling without having these questions in mind, one might be at a high risk to miss the anomaly in question. Some geologists are tempted to sample anything loose within a river valley. Before establishing a reliable locality for stream sediment sampling, another key question to ask is why we need to sample stream sediments? In sampling sediments one needs to pick ore signatures. The ore signatures can be reflected by anomalous values or by other nonconventional techniques as stated in Chapter 3, and the application of nonconventional mineral exploration techniques—case studies, are detailed in Chapter 4. These sediments are therefore coming from an ore body (or multiple ore bodies) or had contacts with fluids (hydrothermal) from the ore bodies. Thus both professionalism and creativeness in planning where to sample are needed.

The first stage is to establish the target to be explored. Second is to find out if there are any stream channels coming from that target area and mark them. The third aspect is to establish a catchment area/-catchment basin, that is, the possible zone(s) where the anomaly can be trapped or fenced optimally. In other words, explorers try to predict the possible directions/and distances the anomaly could have moved from and its trails within that zone (e.g., Nezhad et al., 2017; Perks et al., 2017 and Fig. 1.11).

1.1.3.3 *Termite mounds*

Termite mounds are convoluted features but well-ventilated structures built by termites as their homes (Fig. 1.12). They are built anywhere within weathered profiles of rocks (i.e., on soils, regolith, etc.). Inside the mound is an extensive system of tunnels and conduits that serves as a ventilation system for the underground nest. In order to get good ventilation, the termites construct several shafts leading down a cellar located beneath the nest. Termite mounds rein a variety of shapes and sizes. For instance, Odontotermes termites build open chimneys or vent holes into their mounds while others such as Macrotermes build completely enclosed mounds. The Amitermes (magnetic termites) mounds are created tall, thin, wedge-shaped, usually oriented north–south (Ball, 2010).

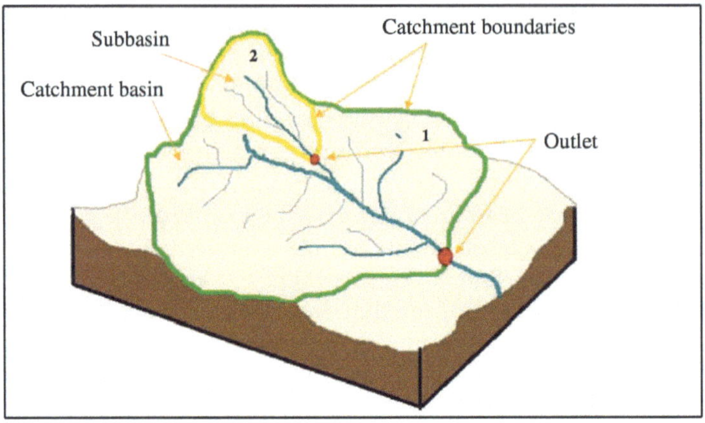

FIGURE 1.11 A block diagram showing a catchment basin, subbasin, catchment boundary, and the outlet. Source: *After Nezhad, G.S., Mokhtari, A.R., Roshani, R.P., 2017. The true sample catchment basin approach in the analysis of stream sediment geochemical data. Ore Geol. Rev. 83, 127–134.*

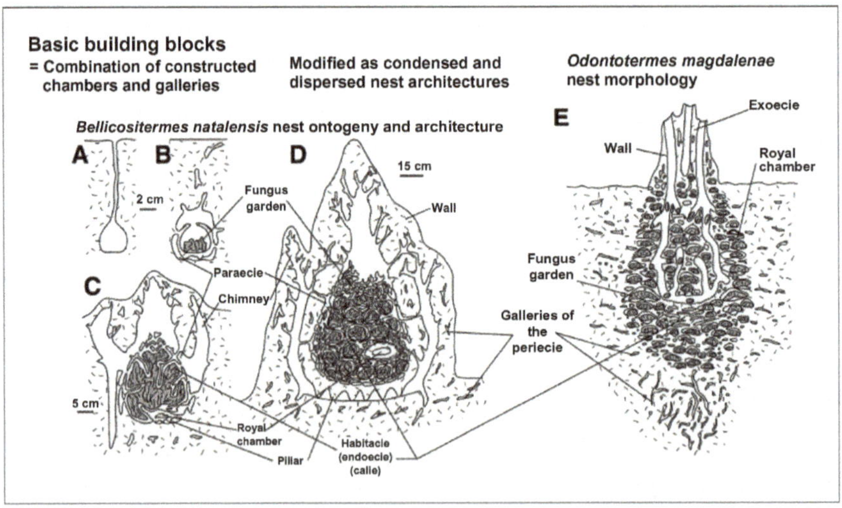

FIGURE 1.12 Examples of extant termite nest morphologies. The basic building blocks of termite nests are constructed chambers and galleries arranged in dilute or concentrated architectures. (A–D) Nest ontogeny and architecture of *Bellicositermes natalensis*. (E) Nest morphology of *Odontotermes magdalenae*. Source: *After Hasiotis, S.T., 2003. Complex ichnofossils of solitary and social soil organisms: understanding their evolution and roles in terrestrial paleoecosystems. Palaeogeogr. Palaeoclimatol. Palaeoecol. 192 (1–4), 259–320. Available from: https://doi.org/10.1016/s0031-0182(02)00689-2.*

Termite mounds are a good sampling media in mineral exploration and have been used extensively in recent years in search for hidden mineral deposits (e.g., Howarth and Govett, 1983). Most importantly, termite mounds show both higher values of most elements, their contrasts, and they also show positive correlation of the elements of similar geochemical behavior (Govett, 1978, 1983; Howarth and Govett, 1983). In northern Ethiopia, termite mounds and the parent materials are positively correlated for Au ($r = 0.75$), Cu ($r = 0.77$), Ag ($r = 0.56$), and Mn ($r = 0.72$); implying direct relation between concentration of metals in termite mounds and the parent rocks (e.g., Govett, 1978; Howarth and Govett, 1983; Nude et al., 2014).

1.1.3.4 *Botanic materials*

Botanic/biological materials refer to vegetation or plants that can be used for mineral prospecting. Such exploration methods are called biogeochemical methods that involve analysis of trace elements in plants to determine ore deposits anomalies. Elements tend to accumulate in plants and upper humic layer of soils and these can be good indicators of ore deposits (Dunn, 2007; Cannon, 1960).

Botanic materials can be of great use in mineral exploration even though different plant species (tissues) have different element concentrations (Table 1.4) and attempts to average them has been a challenge over the decade ago (middlemost, 1994). Copper flowers in Zambia have been extensively used to search for porphyry Cu deposits. Similarly, Calamine violet has led to the discovery of Pb, Zn deposits of West and Eastern Europe (Gandhi and Sarkar, 2016). Thus by analyzing the chemistry of the soil on which Cu flowers or Calamine violet plants grow; explorers are able to recognize some abnormal element concentrations.

1.1.4 Geochemical elements

The previous sections briefly discussed the circulation of elements in nature, particularly within rocks, soils (and regolith materials), plants, and water. When rocks are broken down as a result of weathering and erosion, soils and regolith materials are produced. Rocks, soils, and regolith materials are made up of minerals; and minerals themselves are a function of elements. The next section 1.1.4.1 therefore is devoted to elements: major and trace elements, their abundance, classification schemes, and their significance in mineral exploration.

1.1.4.1 *The periodic table*

In order to appreciate the varsity geological processes, understanding the Periodic Table is necessary. The Periodic Table is a foundation

TABLE 1.4 Element abundances in plants (dry weight)—summary of estimates of worldwide averages of all tissues from all plants.

Elements	Units	Concentration	Elements	Units	Concentration
Major elements (>0.1%)			*Trace elements (<1000 ppm)*		
C	%	44.5	La	pppm	0.2
O	%	42.5	Li	ppm	0.2
H	%	6.5	Lu	ppb	3
N	%	2.5	Mn	ppm	200
K	%	1.9	Mo	ppm	0.5
Ca	%	1	Na	ppm	150
S	%	0.3	Nb	ppb	50
P	%	0.2	Nd	ppm	0.2
Mg	%	0.2	Ni	ppm	1.5
C1	%	0.2	Os	ppb	0.0015
Si	%	0.1	Pa	ppb	?
Trace elements (<1000 ppm)			Pb	ppm	1
Ag[a]	ppb	20	Pd[a]	ppb	0.1
Al	ppm	80	Po	ppb	?
As	ppm	0.1	Pr	ppb	50
Au[a]	ppb	0.2	Pt	ppb	0.005
B	ppm	40	Ra	ppb	?
Ba	ppm	40	Rb	ppm	50
Be	ppb	1	Re[a]	ppb	0.1
Bi	ppb	10	Rh[a]	ppb	0.01
Br	ppm	4	Ru	ppb	0.1
Cd	ppb	50	Sb	ppm	0.1
Ce	ppm	0.5	Sc	ppb	20
Co	ppm	0.2	Se	ppb	20
Cr	ppm	1.5	Sm	ppb	40
Cs	ppm	0.2	Sn	ppm	0.2
Cu	ppm	10	Sr	ppm	50
Dy	ppb	30	Ta	ppb	1

(Continued)

TABLE 1.4 (Continued)

Elements	Units	Concentration	Elements	Units	Concentration
Er	ppb	20	Tb	ppb	8
Eu	ppb	8	Te[a]	ppb	20
F	ppm	2	Th	ppb	5
Fe	ppm	150	Ti	ppm	5
Ga	ppm	0.1	T1	ppb	20
Gd	ppb	40	Tm	ppb	4
Ge	ppb	10	U	ppb	10
Hf	ppb	50	V	ppm	0.5
Hg[a]	ppb	20	W	ppm	0.2
Ho	ppb	8	Y	ppm	0.2
I	ppm	3	Yb	ppb	20
In	ppb	1	Zn	ppm	50
Ir[a]	ppb	0.01	Zr	ppm	0.1

[a]*Indicates modifications by Dunn (2007).*

After Dunn, C.E., 2007. Biogeochemistry in mineral exploration. In: M. Hale (Ed.), The Hand Book of Exploration and Environmental Geochemistry 9, M. Hale (Series editor), Elsevier, Amsterdam, 464 pp; Modified from Markert, B., 1994. Progress report on the element concentrations cadastre project (ECCP) of INTERCOL/IUBS, International Union of Biological Sciences, 25th General Assembly, Paris, 54 pp.

toward understanding chemistry, geochemistry, and earth sciences in general. Many scholars appreciate science because they have a strong foundation on elements and their classification. It is almost impossible to master the world of earth sciences without having a solid base in element types, their classifications, and properties, all which require the background of the Periodic Table.

The Periodic Table, first published in 1869 by Dmitri Ivanovich Mendeleev, a Russian chemist and inventor of the Periodic Law, created a farsighted version of the Periodic Table of elements (Epiotis and Henze, 2003). Dmitri used it to correct the properties of some already discovered elements and to predict the properties of eight elements yet to be discovered. Antoine-Laurent de Lavoisier and John Newlands are other scientists who earlier worked on the Periodic Table. The Periodic Table is the table containing rows and columns of elements that are arranged according to their increasing atomic numbers and atomic weights in rows (Periods) and down columns (Groups) (Fig. 1.13). By moving down a given Group or along a given Period, elements increase in sizes, though the rate of increase in element sizes down Groups is higher than that along Periods.

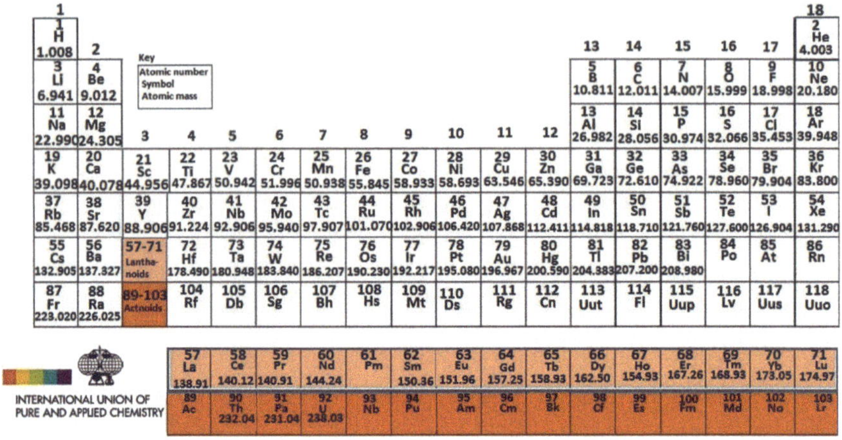

FIGURE 1.13 Periodic Table of elements. The top numbers 1 − 18 represent Group 1 to Group 18 (Gaffney and Marley, 2018).

Dmitri Mendeleev who published the popular Periodic Table in 1869

Elements in the same Group tend to show similar patterns in atomic radius, ionization energy, and electronegativity. From top to bottom in a Group, the atomic radii of the elements increase (Fig. 1.13). Since there are more filled energy levels, valence electrons are found farther from the nucleus. From the top, each successive element has lower ionization energy because it is easier to remove an electron since the atoms are less tightly bound. Similarly each group of elements in the Periodic Table is characterized by a top-to-bottom decrease in electronegativity (electronegativity) due to an increasing distance between valence electrons and the nucleus. There are exceptions to these trends: for example, in Group 11, electronegativity increases farther down the Group (Van Spronsen, 1969).

Elements are abbreviated in the Periodic Table by a letter or two that is/are related to the name of the element itself. For example, V is the symbol for vanadium and Sc for scandium. At times, the symbols do not match the names of elements. For example, Au stands for gold, Hg for mercury, and so on. In such cases, the symbol might be coming from Greek or Latin languages. Some elements are named after scientists who discovered them, such as Lise Meitner (meiterium, Mt). Some named after geographic locations where they were found such as France (francium, Fr).

Newly synthesized elements are given a temporary name and three-letter symbol that is related to the element's atomic number. The International Union of Pure and Applied Chemistry (IUPAC) adopted this system in 1978. Once the discovery of the element is verified, the discoverers can choose a permanent name. Unverified elements to date are those with the atomic numbers 113, 115, 117, and 118 (Fig. 1.13). For example, in June 2016 IUPAC announced the names and symbols of the new four elements in the Periodic Table (previously temporary elements). Elements 113 (Uut) as nihonium (Nh), 115 (Uup) as moscovium (Mc), 117 (Uus) as tennessine (Ts), and 118 (Uuo) as oganesson (Og) (Makishima, 2017). Elements Ms and Ts are names in honor of the cities Moscow and Tennessee following the European–American–Russian Research collaborations. Japan's RIKEN research institution had the honor of selecting the name and symbol of element 113, (Nh) and was the first time Asian researchers have named an element (Makishima, 2017). See also Fig. 2.14 for comparison on newly discovered elements.

1.1.4.2 Abundance of elements in the planet earth and in the solar system

There are about 115 elements on earth as of today. Each of these elements occurs at different concentrations on the earth's compartments: the atmosphere, the biosphere, the hydrosphere, the crust, and the lithosphere and in the solar system (Fig. 1.14). The element concentrations are also called element abundances. Table 1.5 shows average (AVE), upper continental crust (UCC), several published median (MED), global soil (GS), and world soil (WS) reference values, average or median values from continental-scale geochemical survey data from the United States and China (as described in the footnote) compared to the new median soil total concentrations determined by continental-scale geochemical surveys Europe (GEMAS Ap) and Australia (NGSA Tc) (De Caritat and Reimann, 2012).

1.1.4.3 Classification of elements

For the interest of mineral exploration, only the abundances of elements; major and trace elements on the crust (upper crust) shall be

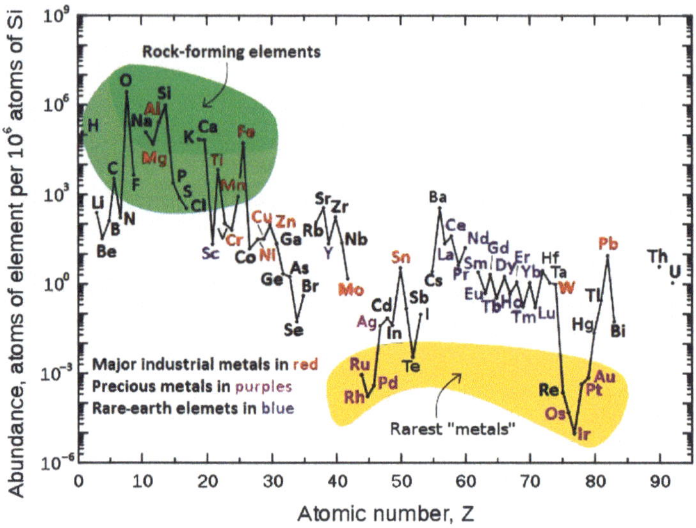

FIGURE 1.14 Relative abundance of the chemical elements in the solar system, expressed as atoms of each element per million atoms of silicon (Si). Elements with atomic numbers 43, 61, 84–89, and 91 have no stable or long-lived isotopes, and therefore have vanishingly small abundances. Source: *After Anders, E., Ebihara, M., 1982. Solar-system abundances of the elements. Geochim. Cosmochim. Acta 46 (11), 2363–2380; Makishima, A., 2017. Origin of Elements. Origins of the Earth, Moon, and Life, pp. 1–24.*

discussed hereunder. Abundance of an element is basically its average concentration in a given earth's compartment (atmosphere, soil, crust, mantle, or core)—for our case, "earth's crust." The crust is made up of different rock types; sedimentary, metamorphic, and igneous rocks. All of these rocks occur at a given locality as a function of earth's activity in space and time. During the Precambrian time, for example, especially during the Hadean and Achaean era, the crust was characterized by essentially mafic/and ultramafic rocks because the crust had not differentiated enough into a think felsic crust like we have today. Conversely, the areas occupied by felsic rocks at present make the upper crust felsic in composition. However, this is a general approximation as some parts of the upper crust have ultramafic rocks such as dunites, komatiites, peridotites, and some parts have almost exclusively felsic rocks whereas some parts of the crust have layered intrusions: ultramafic–mafic–felsic. The soils from either of the rock units will reflect the mineralogical composition of the parent rocks especially if the soil is in situ (residual soils).

1.1.4.3.1 Major elements

Major elements, as the name entails, are those elements that make up large proportion of the earth or solar system, typically more than

TABLE 1.5 UCC: Upper continental crust, from Rudnick and Gao (2003); GS: Global soil, from Vinogradov (1954); WS1: World soil, from Bowen (1979); WS2: World soil, from Koljonen (1992); WS3: World soil, from Kabata-Pendias (2001); USA: Conterminous US soils, AVE from Shacklette and Boerngen (1984), and MED from Garrett (2009) or Gustavsson et al. (2001); China: Chinese soil, from Li and Wu (1999); and Ap, Tc are from De Caritat and Reimann (2012).

	UCC AVE	GS MED	WS1 MED	WS2 MED	WS3 MED	USA AVE	USA MED	China AVE	AP MED	TC MED
Majors total (wt.%)										
Al_2O_3	14.5	13.5	15.1	13.4		13.6	9.7	12.3	10.5	8.1
CaO	3.6	1.9	2.0	2.1		3.4	1.4	3.1	1.2	0.5
Fe_2O_3	5.6	5.4	50.	5.7		3.7	2.8	4.4	3.6	3.2
K_2O	2.8	1.6	1.7	1.7		1.8	1.8	2.3	1.9	2.0
MgO	2.5	1.0	1.5	0.8		1.5	1.0	1.5	1.0	0.5
MnO	0.10	0.11	0.07	0.13	0.07	0.07	0.05	0.08	0.08	0.04
Na_2O	3.27	0.87	1.35	0.67		1.62	1.10	1.52	0.79	0.30
P_2O_5	0.15	0.18	0.17	0.18		0.10		0.15	0.18	0.06
SiO_2	66.6	70.6	59.9	70.6		66.3			66.8	77.5
TiO_2	0.64	0.77	0.67	0.83		0.48	0.42	0.68	0.62	0.58
Trace total (mg/kg)										
As	5	5	5	6	5	7	6		7	3
Ba	624	500	500	500	362	580	502		391	315
Ce	63	(50)	65	50	49	75			59	42
Co	17	8	10	8	7	9	7	13	9	8
Cr	92	200	80	70	42	54	50	58	64	48
Ga	18			20	12	17	15		12	10
Nb	12		12	10	12	11			13	9
Ni	47	40	20	50	18	19	15	25	21	15
Pb	17	10	17	35	25	19	17	25	21	13
Rb	84	60	65	150	50	67			75	51
Sr	320	300	240	250	147	240	148	186	102	68
Th	11	6	9	9	8	9			9	8
V	97	100	90	90	60	80	67	78	70	55
Y	21		20	40	12	25	23		28	21
Zn	67	50	70	70	62	60	52	69	62	31
Zr	193	300	230	400	300	230	188		263	304

TABLE 1.6 Major elements in wt.% ([a]LOI).

SiO_2
Al_2O_3
MgO
CaO
Na_2O
K_2O
Fe_2O_3
TiO_2
MnO
Cr_2O_3
NiO
P_2O_5
LOI[a]

[a]LOI (loss of ignition) is commonly used to measure the volatile species, unburned carbon and moisture in the solid materials and is calculated as LOI through Eq. (1.1).

98.5 wt.% of the total elements, either on earth or in the solar system (Rudnick and Gao, 2003). Major elements are expressed as oxide wt.% (Table 1.6) and add up to about 100% or between 98.5% and 101%.

$$\text{LOI\%} = \frac{\text{Mass loss of sample}}{\text{Mass of original sample}} \times 100 = \frac{W1 - W2}{W1 - W0} \times 100 \qquad (1.1)$$

where $W0$, $W1$, and $W2$ are the masses of blank crucible, crucible with sample before and after ignition, respectively (Mu et al., 2017).

1.1.4.3.2 Significance of major elements in geology

As stated in Section 1.1.4.3.1, major elements make up most of the earth's crust and indeed, the solar system. They also make the rock-forming minerals that include silicates (olivine, feldspars, micas, quartz, pyroxenes, hornblende, etc.), sulfides (pyrites, chalcopyrites, pentlandite, pyrrhotite, etc.), oxides (magnetites, chromites, etc.), and carbonates (e.g., Anders and Ebihara, 1982; Barrat et al., 2012). Therefore the intrinsic properties of rocks are contained within major elements, thus major elements can therefore be used to classify rocks (Figs. 1.3 and 1.5; Table 1.6).

TABLE 1.7 Goldschmidt's (1888–1947) classification of the elements (Goldschmidt, 1954)

Lithophile	Siderophiles	Chalcophile	Atmophile
Li, Na, K, Rb, Cs	Fe^a, Co^a, Ni^a	(Cu), Ag^b	(H), N, (O)
He, Ne, Ar, Kr, Xe	Ru, Rh, Pd, Zn, Cd, Hg^c,	Be, Mg, Ca, Sr, Ba	
B, Al, Sc, Y, REE	Os, Ir, Pt	Ga, In, Tl	
Si, Ti, Zr, Hf, Th	Au, Re^d, Mo^d	(Ge), (Sn), Pb	
P, V, Nb, Ta	Ge^a, Sn^a, W^e	(As), (Sb), Bi	
O, Cr, U	C^e, Cu^a, Ga^a	S, Se, Te	
H, F, Cl, Br, I	Ge^a, As^c, Sb^d	(Fe), Mo, (Os)	
(Fe), Mn, (Zn), (Ga)		(Ru), (Rh), (Pd)	

[a]*Chalcophile and lithophile in the earth's crust.*
[b]*Biophile as well.*
[c]*Chalcophile in the earth's crust and biophile as well.*
[d]*Chalcophile in the earth's crust.*
[e]*Lithophile in the earth's crust.*
After Hollabaugh, C.L., 2007. Modification of Goldschmidt's geochemical classification of the elements to include arsenic, mercury, and lead as biophile elements. Develop. Environ. Sci. 9–31; Makishima, A., 2017. Origin of elements. In: Origins of the Earth, Moon, and Life, pp. 1–24.

1.1.4.3.3 Trace elements

Unlike major elements, trace elements account for <0.1% of the total elements in the earth's crust or in the solar system (Turekian and Wedepohl, 1961). Trace elements include Ni, Cu, Zn, Pb, W, Ta, and so on. Trace elements can be subdivided into several groups namely lithophiles, siderophiles, chalcophiles, atmophiles, and biophiles (Hollabaugh, 2007; Table 1.7). Trace elements can also be further subdivided into platinum-group elements (PGEs)/metals and rare-earth elements (REEs) as is described in Sections 1.3.3.3.2 and 1.3.3.3.3.

1.1.4.3.3.1 Lithophiles, siderophiles, chalcophiles, atmophiles, biophiles Lithophiles are elements that have a high affinity to the crust; that is, they like the crust. Siderophiles are elements that have high affinity to iron; that is, they like to associate with iron. Table 1.7 shows the list of lithophiles, siderophiles, chalcophiles, and atmophiles though Goldschmidt included in his classification the biophile elements (Hollabaugh, 2007). Biophile elements are elements that are the most typical in organisms and organic material, or are concentrated in living plants and animals (Bates and Jackson, 1984). Because biophile elements all occur as lithophile and/or chalcophile elements (C, P, S, Cl, Ca, Mg,

K, Na, V, Fe, Mn, and Cu), 60 years latter different authors in the fields of geochemistry and environment dropped biophiles from the list (Hollabaugh, 2007). However, recently As, Pb, and Hg have been proposed to create the biophile group again, because they have great affinity to concentrate in the biosphere (Hollabaugh, 2007).

1.1.4.3.3.2 Platinum-group elements PGEs, consist of Osmium (Os), Iridium (Ir), Ruthenium (Ru), Rhodium (Rh), Platinum (Pt), and Palladium (Pd). They are transition precious metallic elements with similar physical and chemical properties grouped together in the D-block in Groups 8, 9, and 10; and Periods 5, 6, and 7 of the Periodic Table. The first three elements are called Ir-group (IPGE) and the other four elements (including Au), the Pd-group (PPGE). The distribution of PGE and most other chalcophile elements in mafic and ultramafic rocks is controlled predominantly by sulfides. Most of the world's PGE are produced from two types of deposits: PGE-dominated deposits where PGEs are the main product and Ni–Cu sulfide deposits where PGE are the by-product. Thus PGE occur in close associations with Ni–Cu in form of Ni–Cu–PGEs sulfide deposits characterized by high-sulfur disseminations and massive, and low-sulfur disseminations such as is the case at Merensky Reef (Bushveld), Voisey's Bay, Jinchuan, and Duluth (Naldrett, 1981; Haldar, 2018) and Kabanga (Maier, 2005; Macheyeki, 2011).

Both types of deposits are closely associated with layered intrusions (Maier, 2005). The IPGE and PPGE groups behave differently in such a way that the former tend to be compatible during mantle melting whereas the latter are incompatible (Rollinson, 1993).

1.1.4.3.3.3 Rare-earth elements REEs are a collection of 17 chemical elements in the Periodic Table that occur together in nature. They display similar chemical properties. The 17 REEs are lanthanum (La = 57), cerium (Ce = 58), praseodymium (Pr = 59), neodymium (Nd = 60), promethium (Pm = 61), samarium (Sm = 62), europium (Eu = 63), gadolinium (Gd = 64), terbium (Tb = 65), dysprosium (Dy = 66), holmium (Ho = 67), erbium (Er = 68), thulium (Tm = 69), ytterbium (Yb = 70), lutetium (Lu = 71), scandium (Sc = 21), and yttrium (Y = 39). The uses of REE in mineral exploration are detailed in Chapter 2, Types of ore deposits and their origin.

1.1.4.3.3.4 Compatible and incompatible elements When partial melting occurs, some elements preferentially enter the crystal lattice (solid phase) whereas others enter the melt. Those that preferentially enter the crystal lattices are called compatible elements whereas those that do not are called incompatible elements (Spandler et al., 2017).

Compatible elements in one phase are not necessarily compatible in the other phase (Naldrett, 1981; Spandler et al., 2017). Therefore in defining compatibility and incompatibility of elements, it is important to show the mineral phases that are being considered. For instance Sc is compatible in clinopyroxene but not in olivine, Zr is compatible in zircon, and P in apatite but neither is compatible in olivine or pyroxene (Naldrett, 1981). It is also important to note that incompatible elements are preferentially partitioned into silicate melts rather than coexisting minerals (Spandler et al., 2010, 2017).

In other words, an element fractionates (partitions) or makes an uneven distribution of its ions between two competing (equilibrium) phases (e.g., solid phase and liquid phase) as shown in Eq. 1.2:

$$KD = \frac{X i\text{crystal}}{X i\text{melt}} \tag{1.2}$$

where Xi is the mole fraction of element in the solid or liquid phase and KD is the distribution or partition coefficient of an element (e.g., Bédard, 2005; Kessel et al., 2015).

Mafic and metasedimentary crustal materials within the mantle represent chemically fertile components that undergo partial melting at temperatures well below the anhydrous peridotite solidus (Fig. 1.15).

Further, incompatible elements can be subdivided based on the ratio of charge/ionic radius (Z/r) where $Z/r > 2$ are classed as high field strength elements or else if $Z/r < 2$, they are called large ion lithophile elements (Whittaker and Muntus, 1970; Whittaker, 1989).

The term large ion lithophile element or LILE is usually confusing as one school of thought is that LILEs are incompatible elements whereas the other school of thought is that LILEs are a specific subset of incompatible trace elements characterized by large ionic radius to ionic charge ratio (or low "field strength" elements). Initially, Gast (1972) included the cations K, Rb, Sr, Cs, Ba, REE, Th, and U as LILE. Following the Li large radius to charge ratio, Gast (1972) includes it in the list of LILE even though it is small ($Z/r = 0.82$).

Because of the confusion in the literature regarding usage of LILE, it is recommended that the term be restricted to lithophile trace elements having a large ionic radius to charge ratio that have ionic radii greater than those of Ca^{2+} and Na^{1+}, the largest cations common to rock-forming minerals (Kessel et al., 2015). By this definition, LILE are K, Rb, Cs, Sr, Ba, Pb, and Eu^{2+} (Whittaker, 1989; Williams and Fraústo da Silva, 2006). Recent work by Edmonds (2015) show that Lithium and B are light lithophile elements that partition into volcanic minerals and into vapor from silicate melts, making them potential tracers of

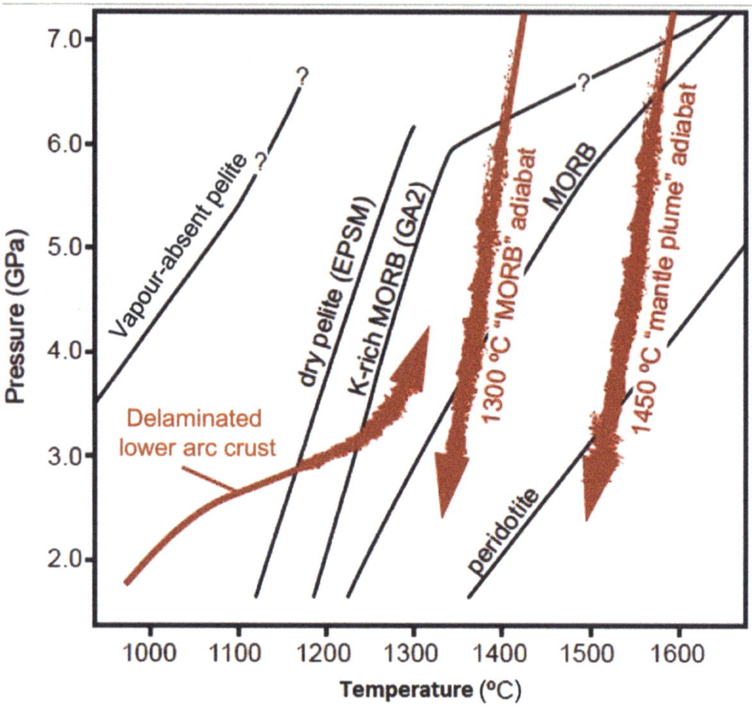

FIGURE 1.15 Goldschmidt's (1888–1947) classification of the elements (Goldschmidt, 1954; Hollabaugh, 2007; Makishima, 2017). Solidus of possible crustal and peridotite components of heterogeneous mantle domains, together with possible mantle adiabatic paths for upwelling mantle beneath an ocean-floor spreading center and for a mantle plume. Also shown is a generalized heating path for delaminated lower crust foundering into the mantle. In all three cases, crustal rocks are expected to undergo significant partial melting. Solidi curves for GA2, EPSM, and the vapor absent pelite are taken from Spandler et al. (2008), Spandler et al. (2010), and Mann and Schmidt (2015), respectively. The solidus curve for peridotite is taken from Hirschmann (2000). Source: *From Spandler, C., Hammerli, J., Yaxley, G.M., 2017. An experimental study of trace element distribution during partial melting of mantle heterogeneities. Chem. Geol. 462, 74–87.*

degassing processes during magma ascent to the surface of earth and of other planets (Whittaker, 1967; Whittaker and Muntus, 1970; Whittaker, 1989).

1.1.4.4 Elements dispersion in primary and secondary environments

In any ore body, the trace elements within its minerals are never static; they are always in a dynamic state. They move from their original (primary/parent) rocks to other parts (secondary) of the ore, rock, or other materials in response to changes of natural parameters such as Eh, pH, oxygen fugacity, temperature, chemistry, and pressure connected to the ore body. The process of trace-element movements is termed as dispersion; it is largely a weathering effect of the ore body (Gandhi and Sarkar, 2016).

If the elements movements are restricted within the area where elements were formed, it is called "primary dispersion." In this case, the size and shape of resulting dispersion halos differ significantly as an aftereffect of the various physical and chemical variables that influence passage of fluid in rocks (Kyser, 2016). Contrarily, in case, trace elements of ore bodies and their associated primary halos are leached away by weathering processes to soils, overburden and vegetation they consequently generate secondary halos (secondary dispersion). Some chemical constituents of ore bodies may be widely dispersed through the agency of ground waters or surface stream systems; analysis of spring and stream waters or stream sediments may therefore indicate the presence of a mineral deposit from a considerable distance (Robb, 2005; Gandhi and Sarkar, 2016; Kyser, 2016).

The trace elements do not disperse in the same rate. Some are more mobile than the others and can therefore form trails of their movement (halos) much faster than the other less mobile elements. Thus metallic mineral deposits are surrounded by halos of abnormal trace-element concentrations proximal to mineralized rocks. These abnormal trace-element concentrations can be found within glacial sediments, soils, springs or stream waters, and stream sediments directly (enclosing) or indirectly connected to weathering of mineral deposits (Fig. 1.16).

1.1.5 Stages in mineral exploration

Mineral exploration is the systematic search for minerals or mineral commodities at a given place using mainly geological, geophysical, and geochemical methods. Mineral exploration is usually undertaken at five stages: target generation, target drilling, resource evaluation drilling, feasibility study, and finally mining. There are three stages of actual field activities and two desk work office stages one at the beginning known as the planning stage and a desktop due-diligence evaluation stage often called the feasibility study at the end. The planning stage covers the selection of commodity, type of deposit, exploration methods, and the setting up of an exploration organization (e.g., Moon et al., 2006).

Mineral exploration stages could be summarized as: (1) target generation, (2) target evaluation, (3) target definition, and (4) resource viability determination. In each stage, there are key activities that characterize that particular stage (Table 1.8) and at every stage, applied geochemistry is a common (and main) activity. It is because of its intrinsic

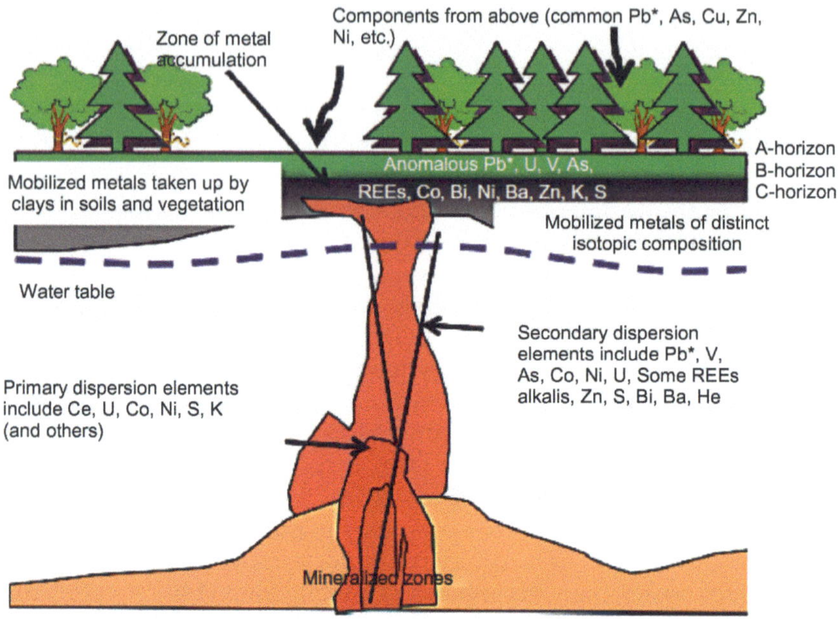

FIGURE 1.16 The diagram of buried unconformity-type uranium deposit showing elements associated with primary dispersion (synmineralization) and secondary dispersion (postmineralization). During secondary dispersion, elements such as REEs are mobilized from the deposit and primary dispersion alteration halo and can be fixed by fracture fillings, clays, and Fe–Mn oxides in soils, and by vegetation. Also shown is the influence of components deposited from above because of anthropogenic activity. Exploration geochemistry targets both primary and secondary dispersion. Source: *After Cameron, E.M., Hamilton, S.M., Leybourne, M.I., Hall, G.E.M., McClenaghan, M.B. 2004. Finding deeply buried deposits using geochemistry. Geochem. Explor. Environ. Anal. 4, 7–32; Kyser, K., 2016. Novel geochemical techniques integrated in exploration for uranium deposits at depth. Uravan Minerals Inc. 26 pp.*

importance that applied geochemistry in mineral exploration (exploration geochemistry) is the key subject of this book (Yuan et al., 2015).

1.1.5.1 Reconnaissance survey (target generation)

Reconnaissance survey stage involves target generation over a vast area normally over 100 km^2. In a reconnaissance survey there are two substages: the planning stage in which most work will be carried out in offices before embarking on the actual reconnaissance survey to evaluate the area and be able to plan the actual exploration. In the planning stage, the exploration crew plans the budget, assesses the geological, topographic and satellite image data and any previous exploration or mining activities in the license area.

TABLE 1.8 Stages of mineral exploration.

Stage	Characteristic activities
Reconnaissance survey (target generation)	Key activities defining this stage include: • Review of all available information on the prospect, such as geological maps, geochemical, geophysical data, and results of previous exploration and known occurrence(s) of minerals • Geological interpretation of air photographs and remote sensed imageries • Regional geological mapping; that includes rocks and sediments grab samples for geochemical testing • Stream sediments geochemical survey analysis and evaluation • Regional airborne geophysical surveys (aeromagnetic, gravity, electromagnetics) • The geological, geochemical, and geophysical anomalous areas identification, and analysis to select targets for detailed exploration
Target evaluation	Key activities defining this stage include: • Detailed localized geological mapping to identify surface topographical features of outcropping or near-surface mineral deposits • Soil and rock geochemical survey, analysis and interpretation • Detailed ground geophysical surveys including; induced polarization (IP), resistivity (R), magnetic and gamma-ray spectrometry • Pitting, trenching, sampling, and pit–trench wall logging to delineate the bedrock geology of anomalous geological, geochemical, or geophysical targets • Preparation of 3D models (e.g., geological models, geochemical models, and geophysical models)
Detailed mineral exploration (target defining—drilling)	Key activities defining this stage include: • Detailed pattern drilling to intersect ore body (s)and understanding the geometry/shape of the ore body as well as grade(s) • Logging and core or chips geochemical analysis for grade determination and resource evaluation • Resource evaluation by calculating ore grade and cutoff grades and tonnages of the minerals contained in the reserve

(*Continued*)

TABLE 1.8 (Continued)

Stage	Characteristic activities
Prefeasibility to feasibility economic evaluation (scoping study—resource viability determination)	Key activities defining this stage include: • A desktop due-diligence study assessment of all factors relevant to that mine—geological, mining, environmental, political, and economic factors • Technical analysis of grade, tonnage, and mining/metallurgical characteristics of the potential ore body • A good understanding of the nature of the mineralization and provides answers on economic questions based on the quantity (detailed) • Determine the confidence of the resources for economic exploitation • Undertake a financial appraisal to decide whether the project should be executed or not

In the actual field activities, a regional geological assessment is undertaken that includes rocks and sediments grab samples for geochemical testing. Reconnaissance regional geochemical stream sediments survey is executed beginning with the determination of the stream pattern of the area that will allow stream sediments sampling program to be executed. The reconnaissance survey also includes regional airborne geophysical surveys [aeromagnetic, airborne gravity and electromagnetic surveys, etc.—depending on the targeted mineral] to recognize anomalous areas. The geological, geochemical, and geophysical anomalous areas are analyzed and compared to determine and select targets evaluation.

1.1.5.2 Mineral exploration—target evaluation

Mineral exploration is the second stage of the key exploration activities to locate surface mineralization and any old workings in a smaller area of identified geological, geochemical, and geophysical anomalies. It is therefore the follow-up of geological, geochemical, and geophysical anomalies from a reconnaissance survey. Again in this stage geological, geochemical, and geophysical methods are employed to follow up the anomalies or targets.

At this stage, pitting, trenching, sampling, and pit–trench wall logging program are conducted to delineate the bedrock geology of the area and obtain geochemical samples in order to confirm the source of geochemical, geological and geophysical anomalies that were found in the reconnaissance survey and determine accurate grade estimates of the deposit. A hand drill or an auger may be used for shallow drilling to understand the geology of the anomaly and ascertain the position of the deposit.

Soil and rock geochemical sampling, analysis, and interpretation are conducted to ascertain or confirm the presence of mineral deposit. The area is gridded and soil samples are taken at fixed intervals to offer a more clear shape and structure of the ore body. At this stage, ground geophysical surveys may be undertaken for conductive ore bodies like sulfide and oxides mineralization. The geophysical methods may include gradient array induced polarization, resistivity, magnetic, and gamma-ray spectrometry (Gadallah and Fisher, 2009).

1.1.5.3 *Detailed exploration (target definition)*

At this stage, detailed pattern drilling is undertaken on the surface expressed ore body; coupled with soil and drill core geochemistry in order to delineate the structures at depth, extent, and grade of the deposit. Geological, geochemical, and geophysical borehole logging are part of key activities to be performed. Three basic types of drilling methods are available: percussion, reverse circulation, diamond core drilling depending on the nature of rocks or sediments covering the deposit. Exploration drilling program will always defined the ore zone and enable mineralization modeling for the reserve estimation and mine planning. Drilling may continue during mining to confirm ore reserves.

1.1.5.4 *Feasibility or scoping study (resource viability determination)*

This is a due-diligence evaluation of a mineral resource after an intensive drilling program is undertaken on an ore deposit to determine the confidence of the resource(s) for economic exploitation. When drilling program is completed, drilling borehole depths, angles of inclination, azimuths of drill holes, average grades for established ore blocks are compiled and worked on using several developed formulae to calculate ore grade, cutoff grades and tonnage of the minerals contained in the reserve. There are well-established evaluation methods including the Kriging methods as described by Krige (1978), Janisch (1986), and Lane (1988).

When the viability of mineral reserve is ascertained, some other factors are assessed including political, economic, and financial viability as well as the environmental evaluation. Finally, mining evaluation is undertaken and financial appraisal undertaken to decide for the project execution or not. The mineral exploration broad stages are summarized in Table 1.8.

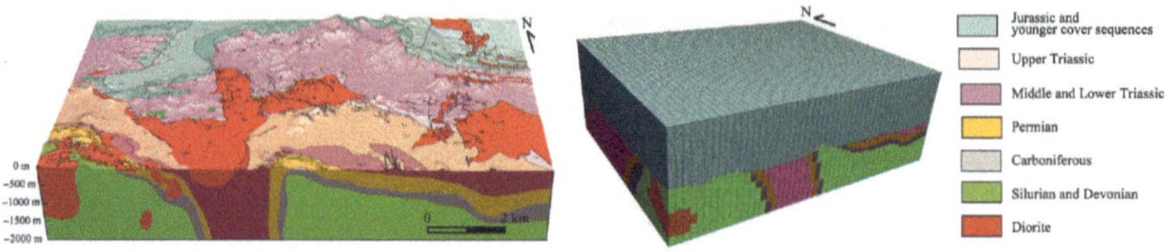

FIGURE 1.17 Example of a 3D geological model prepared at target generation stage. Source: *After Li, X., Young, F., Zhang, M., Jowitt, S., Ord, A., Zhou, T., et al., 2019. 3D computational simulation-based mineral prospectivity modeling for exploration for concealed Fe–Cu skarn-type mineralization within the Yueshanorefield, Anqing district, Anhui Province, China. Ore Geol. Rev. 105, 1–17.*

In mineral exploration, there are geochemical models of element dispersion that can be used right from the onset of exploration geochemistry (i.e., from target generation) in order to anticipate mechanisms of dispersion, to select appropriate sample media, and to estimate the nature and significance of anomalies (e.g., Cohen and Bowell, 2014; Haldar, 2018). Further, the models can assist in identifying blind mineralization in poorly known terrain and in characterizing new styles of mineralization (Cohen and Bowell, 2014; Kyser, 2016) in both primary and secondary environments (e.g., Kyser, 2016; see also Fig. 1.16 on uranium deposits) (Fig. 1.17).

In Chapter 2, Types of ore deposits, their origin, and their characteristics have been described. Further, source of metals in ore deposits have been given. Chapter 3, Conventional and nonconventional exploration techniques—principles, discusses about how to explore various types of ore deposits using conventional and nonconventional exploration techniques. After then, case studies from different parts of the world have been given in Chapter 4. The emphasis is on the use of different exploration techniques, that is, nonconventional exploration techniques including remote sensing geochemistry, concentration gradient modeling, mineral prospectivity modeling and so on.

It is the expectation of the author that this book shall be useful not only to beginners in mineral exploration but also to expert geologists and geoscientists working in mineral exploration and in the mining industry.

References

Anders, E., Ebihara, M., 1982. Solar-system abundances of the elements. Geochim. Cosmochim. Acta 46 (11), 2363–2380.

Azibekov, S.A., Abakumov, E.T., Antropov, P.Y., Alizade, A.A.H., Bagdasarav, V.G., Baryshnikov, A.I., et al., 2019. Mining Encyclopedia (http://mining-enc.com/).

Ball, P., 2010. Termites show the way to the eco-cities of the future. New Sci. 205 (2748), 35–37.

Barrat, J.A., Zanda, B., Moynier, F., Bollinger, C., Liorzou, C., Bayon, G., 2012. Geochemistry of CI chondrites: major and trace elements, and Cu and Zn isotopes. Geochim. Cosmochim. Acta 83, 79–92.

Bates, R.L., Jackson, J.A., 1984. Dictionary of Geological Terms, third ed. American Geological Institute, Anchor Press, Garden City, New York, 571 pp.

Bédard, J.H., 2005. Partitioning coefficients between olivine and silicate melts. Lithos 83 (3–4), 394–419.

Bellieni, G., Visentin, E.J., Zannettin, B., 1995. Use of the chemical TAS diagram (total alkali silica) for classification of plutonic rocks: problems and suggestions. Eur. J. Mineral. 14, 49–52.

Boggs, S., 2006. Principles of Sedimentology and Stratigraphy, fourth ed. Pearson Prentice Hall, 662 pp.

Bowen, H.J.M., 1979. Environmental Chemistry of the Elements. Academic Press, London, 333 pp.

Brady, N.C., Weil, R.R., 1996. The nature and properties of soils, eleventh ed. Prentice Hall Inc, 739 pp.

Brimblecombe, P., 2014. The global sulfur cycle. Treatise Geochem. 559–591. Available from: https://doi.org/10.1016/b978-0-08-095975-7.00814-7.

Cameron, E.M., Hamilton, S.M., Leybourne, M.I., Hall, G.E.M., McClenaghan, M.B., 2004. Finding deeply buried deposits using geochemistry. Geochem. Explor. Environ. Anal. 4, 7–32.

Cannon, H.L., 1960. Botanical prospecting for ore deposits. Sci. Series 132 (3427), 591–598.

Cas, R.A.F., Wright, J.V., 1987. Volcano-clastic Rocks and Successions, Modern and Ancient. Allen and Unwin, London.

Cohen, D.R., Bowell, R.J., 2014. Exploration geochemistry. Treatise Geochem. 623–650.

Coleman, D.C., Callaham, M.A., Crossley, D.A., 2018. Introduction to soil. In: Fundamentals of Soil Ecology, 1–20. Available from: https://doi.org/10.1016/b978-0-12-805251-8.00001-6

Condie, K.C., 2011. Earth's atmosphere, hydrosphere, and biosphere. In: Earth as an Evolving Planetary System, 199–259. Available from: https://doi.org/10.1016/b978-0-12-385227-4.00005-5

Cox, K.G., Bell, J.D., Pankhurst, R.J., 1979. The interpretation of two-element variation diagrams. The Interpretation of Igneous Rocks. Springer, Dordrecht, 468 pp.

De Caritat, P., Reimann, C., 2012. Comparing results from two continental geochemical surveys to world soil composition and deriving Predicted Empirical Global Soil (PEGS2) reference values. Earth Planet Sci. Lett. 319–320, 269–276. Available from: https://doi.org/10.1016/j.epsl.2011.12.033.

Dunham, R.J., 1962. Classification of carbonate rocks according to depositional textures. In: Ham, W.E. (Ed.), Classification of Carbonate Rocks, 1. American Association of Petroleum Geologists Memoirs, pp. 108–121.

Dunn, C.E., 2007. Biogeochemistry in mineral exploration. In: Hale, M. (Ed.), The Hand Book of Exploration and Environmental Geochemistry 9, Hale, M. (Series editor), Elsevier, Amsterdam, 464 pp.

Edmonds, M., 2015. Partitioning of light lithophile elements during basalt eruptions on Earth and application to Martian shergottites. Earth Planet Sci. Lett. 411, 142–150.

Epiotis, N.D., Henze, D.K., 2003. Periodic table (Chemistry). In: Encyclopedia of Physical Science and Technology, 671–695.

Eriksson, P.G., Altermann, W., Nelson, D.R., Mueller, W.U., Catuneanu, O., 2004. Evolution of the hydrosphere and atmosphere. Develop. Precambrian Geol. 359–511. Available from: https://doi.org/10.1016/s0166-2635(04)80007-0.

Fang, Q., Hong, H., Furnes, H., Chorover, J., Luo, Q., Zhao, L., et al., 2019. Surficial weathering of kaolin regolith in a subtropical climate: implications for supergene pedogenesis and bedrock argillization. Geoderma 337, 225–237. Available from: https://doi.org/10.1016/j.geoderma.2018.09.020.

Fitzpatrick, R.W., 2004. Classification systems: Australian. In: Hillel, Daniel (Ed.), Encyclopedia of Soils in the Environment. Elsevier, Oxford, ISBN: 0-12-348530-4, 211–216 pp.

Gadallah, M.R., Fisher, R., 2009. Exploration Geophysics, ninth ed. Springer, 266 pp.

Gaffney, J.S., Marley, N.A., 2018. General chemistry for engineers. Int. Union Pure Appl. Chem. (IUPAC) 623. Available from: https://doi.org/10.1016/b978-0-12-810425-5.09983-5.

Gandhi, S.M., Sarkar, B.C., 2016. Geochemical exploration. In: Essentials of Mineral Exploration and Evaluation, 125–158, Elsevier, 410pp.

Garrett, R.G., 2009. Relative spatial soil geochemical variability along two transects across the United States and Canada. Appl. Geochem. 24, 1405–1415.

Gast, P.W., 1972. The chemical composition and structure of the moon. Moon 5 (1–2), 121–148.

Gillespie, M.R., Styles, M.T., 1999. BGS Rock Classification Scheme. Vol. 1. Classification of igneous rocks. British Geological Survey Research Report, second ed. NERC. RR 99–06.

Goldschmidt, V.M., 1954. Geochemistry. Oxford University Press, Oxford, England, p. 730.

Govett, G.J.S., 1978. Lithogeochemistry, letter to the editor. J. Geochem. Explor. 9, 109–110.

Govett, G.J.S., 1983. Handbook of exploration geochemistry, Rock Geochemistry in Mineral Exploration, third ed. Elsevier, Amsterdam.

Gustavsson, N., Bølviken, B., Smith, D.B., Severson, R.C., 2001. Geochemical landscapes of the conterminous United States—new map presentations for 22 elements. U.S. Geological Survey Professional Paper 1648. 38 pp.

Haldar, S.K. 2018. Exploration Modeling. Mineral Exploration, pp. 195–209.

Haldar, S.K., Tišljar, J., 2013. Metamorphic Rocks. Introduction to Mineralogy and Petrology, first ed Elsevier, 324 pp.

Hartemink, A.E., 2009. The depiction of soil profiles since the late 1700s. Catena 79 (2), 113–127.

Hasiotis, S.T., 2003. Complex ichnofossils of solitary and social soil organisms: understanding their evolution and roles in terrestrial paleoecosystems. Palaeogeogr. Palaeoclimatol. Palaeoecol. 192 (1–4), 259–320. Available from: https://doi.org/10.1016/s0031-0182(02)00689-2.

Hirschmann, M.M., 2000. Mantle solidus: experimental constraints and the effects of peridotite composition. Geochem. Geophys. Geosys. 1. Available from: https://doi.org/10.1029/2000GC000070.

Hoffman, P.F., Kaufman, A.J., Halverson, G.P., Schrag, D.P., 1998. A neoproterozoic snowball earth (PDF). Science 281 (5381), 1342–1346. Available from: https://doi.org/10.1126/science.281.5381.1342. Bibcode: 1998Sci... 281.1342H. PMID 9721097.

Hollabaugh, C.L., 2007. Modification of Goldschmidt's geochemical classification of the elements to include arsenic, mercury, and lead as biophile elements. Develop. Environ. Sci. 5, 9–31.

Howarth, R.J., Govett, G.J.S., 1983. Handbook of exploration geochemistry, Statistics and Data Analysis in Geochemical Prospecting., vol. 2. Elsevier, Amsterdam.

Jackson, J.A., 1997. Glossary of Geology, fourth ed. American Geologic Institute, 769 pp.

Janisch, P.R., 1986. Gold in South Africa. J. S Afr. Inst. Min. Metall. 86, 273–316.

Kabata-Pendias, A., 2001. Trace Elements in Soils and Plants. CRC Press, Boca Raton, FL, 413 pp.

Kafumu, P.D., Paepe, R., 2003. The Quaternary stratigraphy and its associated fossil fauna and flora of the Holili area, NE Tanzania. J. Afr. Earth Sci. 36, 245–250.

Kessel, R., Fumagalli, P., Pettke, T., 2015. The behaviour of incompatible elements during hydrous melting of metasomatized peridotite at 4–6 GPa and 1000 °C–1200 °C. Lithos 236–237, 141–155.

Kisch, H.J., 1979. Mineralogy and petrology of burial diagenesis (burial metamorphism) and incipient metamorphism in clastic rocks. Diagenesis Sediments Sedimentary Rocks 289–493.

Koljonen, T., 1992. Geochemical Atlas of Finland, Part 2 (Ed.): Till. Geological Survey of Finland, Espoo, Finland, 218 pp.

Krige, D.G., 1978. Log Normal—De Wijsian Geostatistics for Ore Evaluation, vol. 1. South African Institution of Mining and Metallurgy Monograph, Johannesburg.

Kyser, K., 2016. Novel Geochemical Techniques Integrated in Exploration for Uranium Deposits at Depth. Uravan Minerals Inc, 26 pp.

Lane, K.F., 1988. The Economic Definition of Ore—Cut-off Grades in Theory and Practice. Mining Journal Books, London.

Lasaga, A.C., Ohmoto, H., 2002. The oxygen geochemical cycle: dynamics and stability. Geochim. Cosmochim. Acta 66 (3), 361–381. Available from: https://doi.org/10.1016/s0016-7037(01)00685-8.

Le Bas, M.J., 1989. Nephelinitic and basanitic rocks. J. Petrol. 30 (5), 1299–1312.

Le Bas, M.J., 2000. IUGS reclassification of the high-Mg and picritic volcanic rocks. J. Petrol. 41 (10), 1467–1470.

Le Bas, M.J., Streckeisen, A.L., 1991. The IUGS systematics of igneous rocks (On behalf of the International Union of Geological Sciences, Sub-commission on the Systematics of Igneous Rocks). J. Geol. Soc. London 148, 825–833.

Le Bas, M.J., Le Maitre, R.W., Streckeisen, A., Zanettin, B., 1986. A chemical classification of volcanic rocks on the total alkali-silica diagram. J. Petrol. 27 (3), 745–750.

Li, J., Wu (Chief Compilers), G., 1999. Atlas of the Ecological Environmental Geochemistry of China. Geological Publishing House, Beijing, China, 209 pp.

Li, X., Young, F., Zhang, M., Jowitt, S., Ord, A., Zhou, T., et al., 2019. 3D computational simulation-based mineral prospectivity modeling for exploration for concealed Fe–Cu skarn-type mineralization within the Yueshanorefield, Anqing district, Anhui Province, China. Ore Geol. Rev. 105, 1–17.

Lyons, T.W., Reinhard, C.T., 2009. Early earth: oxygen for heavy-metal fans. Nature 461 (7261), 179–181. Bibcode: 2009 Natur.461.179L.

Macdonald, E.H., 2007. Gold deposition in the weathering environment. In: Handbook of Gold Exploration and Evaluation, 134–194, Elsevier 664pp. Available from: https://doi.org/10.1533/9781845692544.134.

Macheyeki, A.S., 2011. Application of lithogeochemistry to exploration for Ni-Cu sulfide deposits in the Kabanga Belt, NW Tanzania. J. Afr. Earth Sci. 61, 62–81.

Maier, W.D., 2005. Platinum-group element (PGE) deposits and occurrences: mineralization styles, genetic concepts, and exploration criteria. J. Afr. Earth Sci. 41 (3), 165–191.

Makishima, A., 2017. Origin of elements. In: Origins of the Earth, Moon, and Life, pp. 1–24, Elsevier 268pp.

Mann, U., Schmidt, M.W., 2015. Melting of pelitic sediments at subarc depths: 1. Flux vs. fluid-absent melting and a parameterization of melt productivity. Chem. Geol. 404, 150–167.

Markert, B., 1994. Progress report on the element concentrations cadastre project (ECCP) of INTERCOL/IUBS, International Union of Biological Sciences, 25th General Assembly, Paris, 54 pp.

Middlemost, E.A.K., 1991. Towards a comprehensive classification of igneous rocks and magmas. Earth-Sci. Rev. 31 (2), 73–87. Available from: https://doi.org/10.1016/0012-8252(91)90016-9.

Moon, C.J., Whateley, M.K.G., Evan, A.M., 2006. Introduction to Mineral Exploration, second ed Blackwell Publishing, 481 pp.
Mu, Y., Saffarzadeh, A., Shimaoka, T., 2017. Influence of ignition process on mineral phase transformation in municipal solid waste incineration (MSWI) fly ash: implications for estimating loss-on-ignition (LOI). Waste Manage. 59, 222–228.
Naldrett, A.J., 1981. Platinum-group element deposits. In: Cabri, L.J.(Ed.), Can. Inst. Min. Metall., Spec. Platinum-group Elements: Mineralogy, Geology, Recovery 23, 197–232.
Nezhad, G.S., Mokhtari, A.R., Roshani, R.P., 2017. The true sample catchment basin approach in the analysis of stream sediment geochemical data. Ore Geol. Rev. 83, 127–134.
Nude, P.M., Arhin, E., Yidana, S.M., Foli, G., Dowuona, G.N.N., 2014. Geochemical dispersion of elements and their correlation with gold in the regolith at the Tetteh prospect of the Chirano Gold Mines in the Sefwi belt of the Birimian, southwestern Ghana. J. Environ. Earth Sci. 4 (9), www.iiste.org ISSN 2224-3216 (Paper) ISSN 2225-0948 (Online).
Perks, M.T., Warburton, J., Bracken, L.J., Reaney, S.M., Emery, S.B., Hirst, S., 2017. Use of spatially distributed time-integrated sediment sampling networks and distributed fine sediment modeling to inform catchment management. J. Environ. Manage. 202, 469–478.
Petsch, S.T., 2014. The global oxygen cycle. Treatise Geochem. 437–473. Available from: https://doi.org/10.1016/b978-0-08-095975-7.00811-1.
Pettijohn, F.J., 1983. Sedimentary Rocks, third ed. Harper Collins, 526 pp.
Pope, G.A., 2015. Regolith and weathering (rock decay) in the critical zone. In: Developments in Earth Surface Processes, 113–145, Elsevier 33pp. Available from: https://doi.org/10.1016/b978-0-444-63369-9.00004-5.
Robb, L., 2005. Introduction to Ore-Forming Processes. Blackwell Sciences Ltd, 376 pp.
Rollinson, H.R., 1993. Using Geochemical Data: Longman Geochemistry Series, first ed Routledge, 384 pp.
Rudnick, R.L., Gao, S., 2003. Composition of the continental crust. In: Rudnick, R.L. (Ed.), Treatise on Geochemistry, Vol. 3: The Crust. Elsevier, Oxford, pp. 1–64.
Schön, J.H., 2015. Rocks—their classification and general properties. Physical Properties of Rocks. Fundamentals and Principles of Petrophysics, pp. 1–19.
Shacklette, H.T., Boerngen, J.G., 1984. Element concentrations in soils and other surficial materials of the conterminous United States: an account of the concentrations of 50 chemical elements in samples of soils and other regoliths. U.S. Geological Survey Professional Paper 1270. 105.
Shepard, F.P., 1954. Nomenclature based on sand-silt-clay ratios. SEPM. J. Sediment. Res. 24, 151–158.
Spandler, C., Yaxley, G., Green, D.H., Rosenthal, A., 2008. Phase relations and melting of anhydrous K-bearing eclogite from 1200 to 1600 ºC and 3 to 5 GPa. J. Petrol. 49, 771–795.
Spandler, C., Yaxley, G., Green, D.H., Scott, D., 2010. Experimental phase and melting intraplate magmas. Contrib. Mineral. Petrol. 160, 569–589.
Spandler, C., Hammerli, J., Yaxley, G.M., 2017. An experimental study of trace element distribution during partial melting of mantle heterogeneities. Chem. Geol. 462, 74–87.
Streckeisen, A., 1976. To each plutonic rock its proper name. Earth-Sci. Rev. 12 (1), 1–33. Available from: https://doi.org/10.1016/0012-8252(76)90052-0.
Taylor, G., Eggleton, R.A., 2001. Regolith Geology and Geomorphology. John Wiley & Sons Ltd, 4375 pp.
Turekian, K.K., Wedepohl, K.H., 1961. Distribution of the elements in some major units of the earth's crust. Geol. Soc. Am. Bull. 72 (2), 175.
Van Spronsen, J.W., 1969. The Periodic System of Chemical Elements: A History of the First Hundred Years., first ed. Elsevier, 368 pp.

Verma, S.P., Torres-Alvarado, I.S., Sotelo-Rodríguez, Z.T., 2002. SINCLAS: standard igneous norm and volcanic rock classification system. Comput. Geosci. 28 (5), 711–715. Available from: https://doi.org/10.1016/s0098-3004(01)00087-5.

Vinogradov, A.P., 1954. Geochemie Seltener und nur in Spuren Vorhandener Chemischer Elementeim Boden. Akademie Verlag, Berlin, 249 pp.

Wentworth, C.K., 1922. A scale of grade and class terms for clastic sediments. J. Geol. 30 (5), 377–392. vol. 30, no. 5 (Jul.–Aug., 1922), pp. 377–392 (16 pages), The University of Chicago Press.

Whittaker, E.J.W., 1967. Factors affecting element ratios in the crystallization of minerals. Geochim. Cosmochim. Acta 31 (12), 2275–2288. Available from: https://doi.org/10.1016/0016-7037(67)90002-6.

Whittaker, E.T., 1989. A Treatise on the Analytical Dynamics of Particles and Rigid, fourth ed. Cambridge Mathematical Library, 480 pp.

Whittaker, E.J.W., Muntus, R. 1970. Ionic radii for use in geochemistry. Geochim. Cosmochim. Acta 34, 945–956.

Williams, R.J.P., Fraústo da Silva, J.J.R., 2006. The evolution of earth—the geochemical partner of the global ecosystem (5 billion years of history). Chem. Evol. 1–34. Available from: https://doi.org/10.1016/b978-044452115-6/50044-0.

Yuan, F., Li, X., Zhou, T., Deng, Y., Zhang, D., Xu, C., et al., 2015. Multifractal modelling-based mapping and identification of geochemical anomalies associated with Cu and Au mineralization in the NW Junggar area of northern Xinjiang Province, China.

Further reading

Clement, C.R., 1979. The origin and infilling of kimberlite pipes. In: Kimberlite Symposium II, Cambridge (Extended abstracts).

Chernyshenko, S.V., 2008. Matter and matter flows in the biosphere. In: Encyclopedia of Ecology, 2279–2292. Available from: https://doi.org/10.1016/b978-008045405-4.00743-6.

Fisher, R.V., Smith, G.A., 1991. Volcanism, tectonics and sedimentation, 1–5. In: Sedimentation in Volcanic Settings. SEPM (Society for Sedimentary Geology) Special Publication 45.

Haldar, S.K., 2017. Exploration guide. Platinum-Nickel-Chromium Deposits, pp. 247–266.

Jacob, D.J., 1999. Introduction to Atmospheric Chemistry. Princeton University Press, 279 pp.

Roy, R., Cassard, D., Cobbold, P.R., Rossello, E.A., Billa, M., Bailly, L., et al., 2006. Predictive mapping for copper–gold magmatic-hydrothermal systems in NW Argentina: use of a regional-scale GIS, application of an expert-guided data-driven approach, and comparison with results from a continental-scale GIS. Ore Geol. Rev. 29, 260–286.

White, W.M., 2013. Geochemistry, first ed. Wiley-Blackwell, 668 pp.

CHAPTER

2

Types of ore deposits and their origin

2.1 Introduction

The crustal abundance of geochemical elements that form ore deposits containing valuable metals that can be explored for economic gain is examined in the following sections in terms of their occurrences and their respective origins, genesis and classification.

An ore is a metalliferous mineral or an aggregate of metalliferrous minerals, more or less mixed with gangue, which from the standpoint of (1) a miner, can be won at a profit, or from the standpoint of (2) a metallurgist, can be treated at a profit (Evans, 1993). On the geological point of view, mineral deposits are geochemical anomalies in the earth's crust due to their local high concentration of one or a diverse suite of trace elements [Ni, Au, Ag, Co, Cr, Sc, rare earth element (REE), platinum group elements (PGEs), high field strength elements (HFS), etc.], that may develop geochemical footprints (González-Álvarez et al., 2016). Economically, mineable aggregates of ore minerals are termed as ore bodies, ore shoots, ore deposits, or ore reserves.

There is almost every mineral at any place in or on the crust at different quantities; that is, at high, low, or trace concentrations. The minerals that occur in higher concentrations in many localities are known as abundant minerals and those in lower proportions, scarce minerals. Similarly, metals that occur in higher concentrations and lower concentrations are known as abundant and scarce metals, respectively.

Most abundant metals and minerals do not attract the attention of mineral explorers, whereas those that occur in trace amounts are of interest. Metals like gold, silver, copper, nickel, etc occur rarely in the crust, while at the same time, their uses are highly needed—just in

Applied Geochemistry
DOI: https://doi.org/10.1016/B978-0-12-819495-9.00002-5

© 2020 Elsevier Inc. All rights reserved.

accordance to the law of supply and demand: the higher the supply, the lower the demand and vice versa. Geochemical abundance and scarcity of metals when plotted using crustal abundances against production estimates allow estimates to be made of the relative rates of depletion of certain metals relatives to others (e.g., Einaudi, 2000; Laznicka, 2014; Peck and Huminicki, 2016).

Metals used in industrial and technological applications can be divided into two classes based on their abundance in the earth's crust. The geochemically abundant metals, of which there are five (Al, Fe, Mg, Mn, and Ti), constitute $>0.1\%$ by weight of the earth's crust, while the geochemically scarce metals, which embrace all other metals (including familiar ones such as Cu, Pb, Zn, Au, Ag), constitute $<0.1\%$. In almost every rock, at least tiny amounts of all metals can be detected by sensitive chemical analysis.

2.1.1 Occurrence of ore deposits

Whether minerals can occur in higher or trace quantities (as crustal abundances, Table 2.1), there is always a level at which they can be considered economic or noneconomic. For instance, although gold is present almost everywhere in the crust, economic mineral deposits are isolated. On the other hand, although quartz is almost present everywhere in large quantities, the economic quartz needs to meet some criteria of purity, mineral associations, type, and quantity and yet, the abundance of quartz makes its price inferior to rare metals/minerals such as gold, tanzanite, etc.

It should be noted here that, average crustal values are different from upper or lower crustal values. For instance, the average Clarke value of Cu in the upper crust is 55 ppm (Ridley, 2013) but that of the average crust is about 27 ppm (Heinrich and Candela, 2014; Table 2.2). See also Hu and Gao (2008).

For an economic mineral deposit to form, four factors should be met; (1) presence of source of the ore components (metals and ligands such as S, H, C, Cl, S, or the combination), (2) transportation mechanism of the ore to an appropriate site (fluid such as H_2O, biological and/or mechanical means), (3) depositional mechanism that puts together the ore components into ore minerals, and (4) appropriate geological setting for the ore body to be preserved (e.g., Guilbert and Park, 1986).

Regarding the factors to be considered on the mineralogy of a mineral deposit so as to classify it as economic or not, Moon et al. (2006) considered the mineralogical form (e.g., native Cu vs chalcopyrite, $CuFeS_2$) and the undesirable constituents (e.g., tennantite, $Cu_{12}As_4S_{13}$ vs copper concentrates) as key factors. They summarize the information

TABLE 2.1 Relative abundances of some elements in the earth's crust.

Element	Abundance (at.%)	Abundance (wt.%)
Oxygen	63	47
Silicon	21	28
Aluminum	6.5	8
Iron	1.9	5
Calcium	1.9	3.6
Sodium	2.6	2.8
Potassium	1.4	2.6
Magnesium	1.8	2.1
Titanium		0.44
Hydrogen		0.14
Manganese		0.1
Phosphorus		0.1
Cu, Cr, Ni, Pb, Zn,		10^{-2}–10^{-3} each
Mn, Sn, U, W		~10^{-4} each
Ag, Hg		~10^{-6} each
Au, Pt		~10^{-7} each

Bailey, R.A., Clark, H.M., Ferris, J.P., Krause, S., Strong, R.L., 2002. The earth's crust. In: Ronald A. Beiley, (ed), Chemistry of the Environment, second edition, pp. 443–482, Academic Press, 835p.

that is required from a sample in order to know whether a given deposit is economic or not includes some, or all, of the following: (1) grade of the economic minerals, (2) the bulk chemical composition, (3) the minerals present, (4) the proportions of each of these minerals and their chemical compositions, (5) their grain size, (6) their textures and mineral locking patterns, and (7) any changes in these features from one part of an ore body to another.

Of the items (1)–(7) listed previously, item (4) on mineral proportions is briefly explained as an example of information useful in determining whether an ore body is economic or not. Consider mineralized systems of PGEs. Strong correlations between Pt/Ti, Pd/Ti, and Rh/Ti ratios indicate mineralized systems of PGE as compared to other variations attributable to magmatic processes, such as olivine accumulation and fractionations that can be depicted from variations of PGE/Ti ratios (Fiorentini et al., 2018). An unusual high concentration of SiO_2, MgO, and Cr and low Al_2O_3 comprise 75% of the world PGE resource,

TABLE 2.2 Typical mineable ore cutoff grades of selected elements in comparison with their average crustal abundance.

Element	Deposit type	Typical ore grade (ppm = gt^{-1})	Concentration in crust (ppm)	Enrichment factor	Largest deposits (example)	Tonnage	Grade
Al	Lateritic bauxite	240,000	84,200	3	Weipa (Australia)	149 Mt	28%
Fe	Banded iron formation upgraded	640,000	52,200	12	Mount whaleback (Australia)	>1800 Mt	65.0%
Ti	Mafic intrusions (ilmenite)	80,000	57,000	14	Lac tio (Canada)		
P	Marine sedimentary deposits	100,000	570	180	Meade Peak member (United States)	>10^{11} Mt	10%
V	Layered mafic intrusions (magnetite)	2000	138	14	Bushveld Complex (South Africa)	50 Mt	0.8%
Cr	Layered mafic intrusions (chromite)	160,000	135	1200	Bushveld Complex (South Africa)	2700 Mt	19%
Zn	Stratiform Pb–Zn–Ag deposits	87,000	72	1200	Broken Hill (Australia)	284 Mt	11%
Ni	Magnetic Ni–Cu–Sulfide	13,000	59	220	Jinchuan, Baijiazuizi (China)	515 Mt	1.1%
Rb	Pegmatites from Rb-bearing Lepidolite	5000	49	100	Tanco (Canada)	0.5 Mt	1.0%
Cu		5000	27	180	El Taniente (Chile)	2850 Mt	1.3%
Cu	Hosted deposited copper sediments	23,000	27	850	Lubin (Poland)	2600 Mt	2.0%
Pb	Stratiform Pb–Zn–Ag deposits	47,000	11	4300	Broken Hill (Australia)	160 Mt	2.0%

Nb	Carbonitites	1900	8	240	Seis Lagos (Brazil)	2898 t	2%
Eu	Carbonitites	1300	1.1	1200	Bayan obo (China)	750 Mt	4.1%
W	Skarn replacement, and vein deposits	3300	1	3300	Shizhuayan (Skarn, China)	270 Mt	0.22%
Mo	Porphryry molybdenum deposits	2000	0.8	2500	Climax (Colorado, United States)	769 Mt	0.22%
Ta	Pegmatites	1800	0.7	2600	Tanco (Canada)	2.07 Mt	0.18%
Ag	Polymetallic vein deposits	160	0.056	2900	Cerro Rico de Potosi (Bolivia)	985 Mt	166 gt^{-1}
In	Volcanic-hosted massive sulfide deposits	24	0.052	500	Kidd Creek (Canada)	134 Mt	50 gt^{-1}
Hg	Epithermal veins and breccias	10,000	0.030	300,000	Almaden (Spain)	40 Mt	>1%
Pt	Magmatic PGE–Ni–Cu sulfide deposits	3	0.0015	2000	Bushveld Complex (South Africa)	9815 Mt	2.3 gt^{-1}
Au	Hydrothermal vein deposits	2	0.0013	1200	Muruntau (Uzbekistan)	2200 Mt	2.4 gt^{-1}
Au	Modified Archean placers	5	0.0013	4000	Witwatersrand Basin (South Africa)	>10^4 Mt	5 gt^{-1}

PGE, Platinum group element.
Tonnage (total mass of mined and economically mineable ore) and average grade of some of the largest ore deposits of each type. All values in weights units. Average concentration in crust from Rudnick and Gao (2003). Estimates of average grades indicative only, mostly based on USGS and Canadian Geological Survey compilations but partly representing single big deposits. Grades and tonnages of some of the largest deposits mostly represent single mines, but in the case of the Meade Peak phosphorite, the Bushveld Complex, and the Witwaters and Basin are estimates for one or several rock units that are mineralized at presently economic ore grades for the respective elements (Heinrich and Candela, 2014).
From Heinrich, C.A., Candela, P.A., 2014. Fluids and ore formation in the Earth's crust. In: Treatise on Geochemistry, pp. 1–28.

Bushveld complex, Great Dyke of Zimbabwe, and Still Water (Naldrett, 2010). The Fe oxide–Cu–Au (IOCG) deposits are commonly associated with presulfide sodic or sodic–calcic alteration and have abundant to low Ti–Fe oxides and/or Fe silicates intimately associated with, but generally paragenetically older than, Fe–Cu sulfides, have light rare earth elements (LREE) enrichment and low S-sulfides [lack of abundant pyrite (Groves et al., 2010)]. According to Macheyeki (2011), Pd/V and (Pd/V)/(Cu/Cr) ratios discriminate between barren and mineralized environments (see Chapter 3: Conventional and nonconventional exploration techniques—principles,and Chapter 4: Application of nonconventional mineral exploration techniques—case studies for details).

2.1.2 Genesis and classification of ore deposits

2.1.2.1 Syngenetic and epigenetic genesis of ore deposits

Ore deposits of economic value are formed under varied geological conditions. Two groups of ore deposits may be established syngenetic and epigenetic depending on their formation time relationship to the rocks associated with them (Robb, 2005). Syngenetic mineral deposits are formed at the same time as the associated rocks as in magmatic segregation during the orthomagmatic stage of consolidation of magma or during precipitation of sedimentary rocks. Formation of ore deposits in gossans and laterites due to in situ residual gossans are also considered as syngenetic because as the new rock (laterite, bauxite, kaolinite, or duricrust) is formed it is at that same time the ore deposit is formed.

The epigenetic mineral deposits are formed later after the enclosing or host rocks have been formed in filled or opened fissures in the country rocks and such ore bodies are called lodes or veins. These vein and lode deposits occur as in interstices of the country where the rock forms first and then ores form as impregnations or replacements of the country rock. In contact metamorphism mineral deposits form irregular ore bodies on the margins of metamorphosed rocks. In sedimentary rocks epigenetic processes, ore deposits are formed due to weathering and deposition of detrital sedimentary rocks in basins where placer deposits are emplaced.

2.1.2.2 Origin and classification

Ore deposits can be classified according to the genesis of the deposits (e.g., Cox and Singer, 1986), host rocks such as shale hosted deposits (Misra, 2000; Robb, 2005; Moon et al., 2006), breccia pipes, the minerals contained within the deposit such as porphyry copper deposits (e.g.,

Moon et al., 2006), and the shape or size of the deposit such as strataform and stratabound deposits (Evans and Moon, 2006).

Basic genetic processes that lead to the concentration of minerals include (1) magmatic mineral deposits that get concentrated in igneous rocks; (2) hydrothermal mineral deposits that form in association with magma and water; (3) sedimentary mineral deposits that are precipitated from a solution, typically seawater; (4) placer minerals sorted and distributed by flow of water (or ice); and (5) residual mineral deposits formed by weathering reactions at the earth's surface (e.g., Moon et al., 2006). Meyer (1981) ore classification is based on both genetic type and rock association. The classification based on both genesis and rock association relates the rock-forming process to ore-forming process, which is more descriptive empirical observation as compared to the genetic linkage alone that tends to have controversy (e.g., Robb, 2005).

Major theories of ore genesis are diverse (Niggli, 1929; Schneider, 1964; Lindgren, 1922, 1933; Bateman, 1950; Guilbert and Park, 1986; Stanton, 1972); in other words, there are no acceptable theories that can explain the entire ore genesis we see today. Stanton (1972) in his account gives much detail on these theories and the reader is referred to that work and the references therein for details. The reader should be cautioned, however, that very often several processes contribute to the formation of an ore body. Thus where we have rising hot aqueous solution forming an epigenetic stockwork deposit just below the surface and passing on upward through it to form a contiguous syngenetic deposit under, say, marine condition, even the above simple classification is facing difficulties. This is the reason why ore geologists, besides producing a plethora of ore genesis theories, have also created a plethora of ore body classification! (e.g., Ridley, 2013).

While in the 19th century, classifications of ore deposits were based on form, texture, and mineral content, Niggli's classification in 1929 was based on volcanic–plutonic rocks and that of Lindgren's classification in 1933 was based on depth temperature. Schneider's classification (1941) was based related to magmatic processes, the latter 20th century to-date classification emphasizes on theories of genesis and environment of deposition (transport medium). In the latter classification, types of ore deposits are presented along with their nature of forming processes.

There are basically three types of ore deposits based on their environment of deposition and genesis: (1) deposits due to internal processes, (2) deposits due to external/surface processes, and (3) deposits due to both internal and surface processes (after Moon, 2016). However, based on ore genetic origin, there are two types of ore deposits: (1) ore deposits formed out of or related to igneous intrusions, such as porphyry-type deposits (PD) and IOCG and (2) sedimentary deposits such as

Carlin-type and Mississippi Valley type (MVT) deposits (Cline et al., 2005; Leach et al., 2005; Seedorff et al., 2005; Williams et al., 2005; Groves et al., 2010; Richards, 2011; Vigneresse, 2019). In this book, classification based on genetic and environment of formation/deposition has been adopted.

2.1.2.3 *Deposits due to internal processes*

Deposits in this group include magmatic (orthomagmatic) ore deposits (Robb, 2005; Moon et al., 2006; Heinrich and Candela, 2014; Moon, 2016). These deposits form as a result of element enrichment from melts at high temperature without essential involvement of aqueous fluids (Heinrich and Candela, 2014). In this type of deposits, heat source (engine) is required in order to drive magmas from different sources in the form of convective cells at a global scale taking fertile magma in the mantle to the crust (e.g., Robb, 2005; Heinrich and Candela, 2014). By so doing, one or several magmatic melt phases are involved that would lead to element partitioning between melts and a phase to be favored is enriched with elements that form the ore deposit (e.g., Robb, 2005; Moon et al., 2006; Heinrich and Candela, 2014). An obvious question would be why some parts of the mantle magma are "fertile?" Some magmas can be more fertile than the other because either those magmas inherited surplus of potential ore-forming trace elements or because their source was already enriched in those components (Robb, 2005). Orthomagmatic deposits can be subdivided into:

1. Deposits formed by crystallization of magma (during differentiation of a silicate melt). Chromium, a lithophile element, is a good example of a mineral that crystallizes from silicate magma as a result of magmatic differentiation (e.g., Ridley, 2013). Chromium is a compatible mineral in spinel and in clinopyroxene crystals relative to ultramafic and mafic melts. Further, chromium can largely be retained in the mantle during partial melting and occurs in significantly lower concentrations in average crust ($\sim$100 ppm) than in mantle (concentrations up to 1%).

   Highest concentration of Cr in the crust occurs within ultramafic layers of large mafic–ultramafic igneous bodies and chromites in ultramafic rocks are solid solutions with varying amounts of Al^{3+} and Fe^{3+} replacing Cr^{3+}, and Mg^{2+} replacing Fe^{2+} (Ridley, 2013). Chromite ores are bodies with greater than about 30% chromite, in many cases almost monomineralic chromitite (=chromite rock), in mafic and ultramafic intrusive rocks. Two common types of chromite deposits are recognized on the basis of ore body form and their geological environment: stratiform chromite deposits in large, layered ultramafic–mafic intrusions; and podiform chromite ores in ophiolites or "Alpine peridotites" (Ridley, 2013). Examples of these types of deposits are the Cr deposits in the Bushveld complex in the Republic of South Africa and the Cr in the Great Dyke of Zimbabwe. Also magma crystallized deposits are exemplified by the V-magnetite, Ilmenite in mafic intrusions (Heinrich and Candela, 2014).

Segregation of sulfide melt from silicate melt (i.e., formation of immiscible sulfide melt phases)—the processes that lead to the formation of Ni–Cu sulfide deposits in mafic and ultramafic rocks. Liquation, liquid immiscibility, settling out from magmas of sulfide, sulfide oxide, or oxide melt that accumulate beneath the silicates or injected into wall rocks or in rare cases erupted on the surface (e.g., Naldrett, 2010; Macheyeki, 2011; Ridley, 2013; Heinrich and Candela, 2014; Moon, 2016). Generally, the Ni–Cu deposits are usually associated with PGE as by-products (Naldrett, 1997, 2004).

2. Deposits formed from immiscible sulfide melt phases: PGE sulfide deposits (Ridley, 2013). Economic PGE primary deposits of this type are laterally extensive as thin layers or reefs, from less than 1 m thick to about 20 m thick in layered ultramafic to mafic large-volume and extensive intrusions in the upper crust. The reefs are broadly parallel to the cumulate layering, and are continuous along much of the extent of the exposed intrusions. There may be multiple reefs at different levels in one layered intrusion. Excluding by- and coproduct PGE resources of Ni–Cu sulfide deposits, almost all known economically significant PGE resources are in three exceptionally large intrusions more than about 4 km thick ($>10{,}000\ km^3$ volume), which have repeated multiple cycles of rhythmic cumulate layering and reversals of composition trends of cumulate minerals of layers of the Great Dyke of Zimbabwe (Naldrett, 1997; Mason-Apps, 1998; Naldrett, 2004; Ridley, 2013).

The segregation of sulfide melt from silicate melt and the immiscibility of sulfide phases are actually promoted by tectonic activities. Both Ni–Cu and PGE, that is, Ni–Cu (PGE) deposits can be well understood and appreciated when looked at a global scale and therefore anticipated that this interplay of tectonics with the dynamic behavior of magmatic systems will become increasingly recognized, and has important implications for deposit exploration (Figs. 2.1 and 2.2; Begg et al., 2018).

3. Deposits formed from low-degree (small fraction) partial melting of mantle (Ridley, 2013; Heinrich and Candela, 2014). The partial melting of a small fraction of a mantle leads to deposition of LREE in carbonatites (e.g., Yang and Santosh, 2015).
4. Deposits formed through the incorporation of a mineral from depth in the earth into magma: diamond deposits in kimberlites and

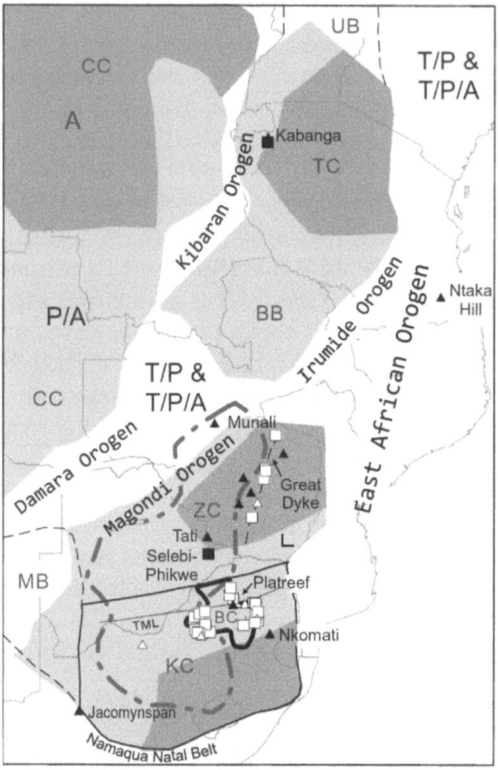

FIGURE 2.1 Deposits and lithospheric resilience. Relationship of giant (squares) and major (triangles) Ni–Cu(-PGE; black) and PGE (white) deposits of Africa with respect to archons (A), proton/archons (P/A), and intervening areas of tecton/proton (T/P) and tecton/proton/archon (T/P/A). Deposits are preferentially located within, or at the margins of the most resilient lithosphere (A, P/A). Thin black lines outline the Kaapvaal Craton and the internal paleocraton boundary represented by the TML. The high-velocity SCLM root at 100–175 km depth (thick gray dash-dot line) is taken from the updated 2011 seismic tomography model of Grand (2002). Note the position of the Great Dyke (dashed line) along the edge of this anomaly, suggestive of a paleocraton boundary. *BB*, Bangwelu block; *BC*, Bushveld complex (thick black outline); *CC*, Congo Craton; *KC*, Kaapvaal Craton; *L*, Limpopo Belt; *MB*, Maltahohe block; *TC*, Tanzania craton; *UB*, Ugandan block; *ZC*, Zimbabwe craton (Begg et al., 2018).

lamproites (Ridley, 2013; Fig. 2.3). It is the process responsible for the formation of diamond in Kimberlites (Republic of South Africa, Canada, Tanzania, etc.). This process also leads to the formation of strongly alkaline silicate igneous rocks such as in the Kola alkaline province of the Russian Federation (e.g., Petrov, 2004; Heinrich and Candela, 2014).

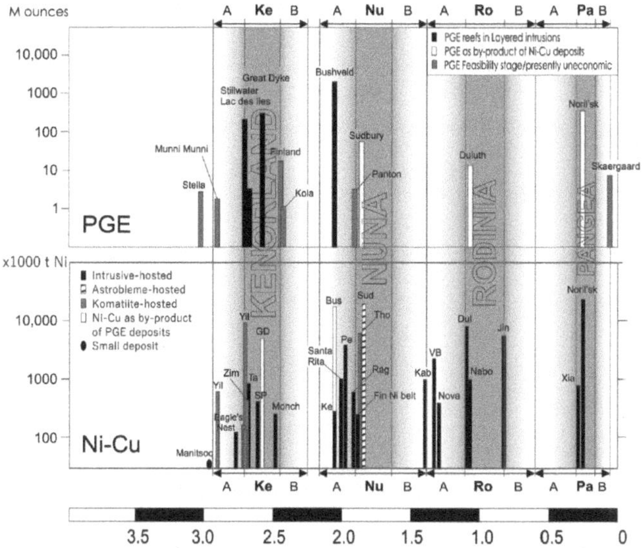

FIGURE 2.2 Secular distribution of PGE and Ni–Cu deposits and key elements of the supercontinent cycle. Supercontinent peaks Kenorland, Ke, Nuna, Nu, Rodinia, Ro, and Pangea, Pa, are preceded by periods of amalgamation, A, and dispersed during periods of breakup, B. Komatiitic-hosted (high-MgO) deposits are restricted to the earlier two supercontinent cycles. Note the clustering of deposits about the amalgamation and peak stages. *Bus*, Bushveld; *Dul*, Duluth; *Fin Ni belt*, Finnish Ni belt; *GD*, Great Dyke; *Jin*, Jinchuan; *Kab*, Kabanga; *Ke*, Keivitsa; *Monch*, Monchegorsk; *Nebo*, Nebo–Babel; *Pe*, Pechenga; *Rag*, Raglan; *SP*, Selebi Phikwe; *Sud*, Sudbury; *Ta*, Tati; *Tho*, Thompson; *VB*, Voisey's Bay; *Xia*, Xinjiang Province deposits (including Huangshandong, Huangshan, Kalatongke, Xiangshan); *Yil*, Yilgarn; *Zim*, Zimbabwe (Begg et al., 2018). Source: *Modified after Maier, W.D., Groves, D.I., 2011. Temporal and spatial controls on the formation of magmatic PGE and Ni-Cu deposits. Mineralium Deposita 46, 841–858.*

These deposits are formed through extreme fractionation of magma: rare-metal pegmatites (Ridley, 2013; Heinrich and Candela, 2014; Dill, 2015, Fig. 2.4). The principle of magmatic fractionation and therefore enrichment/depletion of partial elements obeys the equation (Robb, 2005):

$$\frac{C_{\text{liq}}}{C_{\text{o}}} = \frac{1}{D_{\text{res}} + F(1 - D_{\text{res}})} \tag{2.1}$$

where C_{liq} is the concentration of a trace element in the liquid (melt); C_{o} is the concentration of trace element in parental (unmelted) solid; D_{res} is the bulk partition coefficient of the residual solid (after the melt is extracted); F is the weight fraction of melt produced (Rollinson, 1993, 2012).

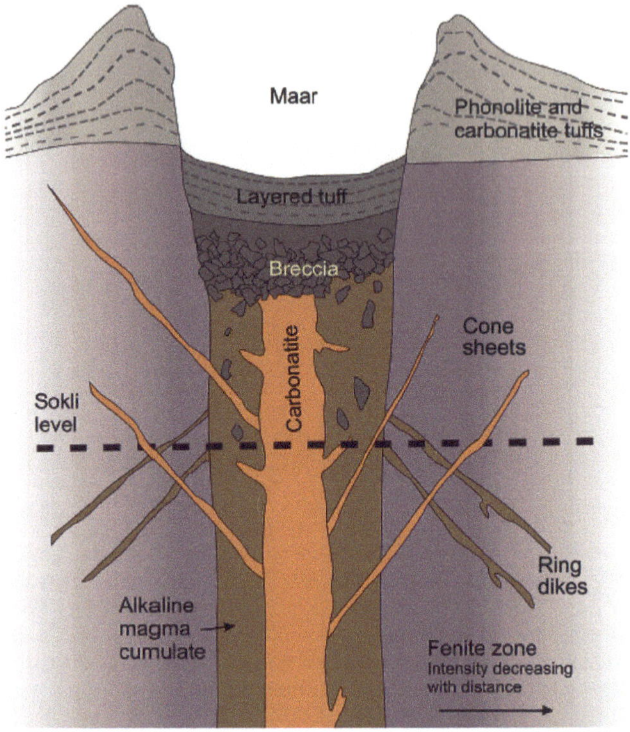

FIGURE 2.3 Cross section of a standard carbonatite, with Sokli erosion level estimated (O'Brien, 2015).

It should be noted that the extraction of a partial melt from its residue, whether it be from an igneous or sedimentary protolith, is a process that segregates chemical components (i.e., fractionation). Partial melts can be considerably enriched in certain elements, but depleted in others, relative to the source rock (Robb, 2005). Pegmatites are enriched with elements that are mostly lithophiles and include the large ion lithophile (LILEs) (Li, Rb, Cs, Be), high field strength elements (HFSEs) (Ga, Sn, Hf, Nb, P, Ta, Y, U, Th, REEs), and strongly soluble elements in aqueous solutions (B, F). Examples of currently or recently exploited pegmatites include: Rössing, Namibia for U (Berning et al., 1976); Tanco, Manitoba, Canada for spodumene, Cs, Ta (Černý et al., 1996); Volta Grande, Brazil; Kenticha, Ethiopia; and Green bushes and Wodgina, Western Australia for Sn, Ta (Partington, 1990).

Hydrothermal deposits are deposits from hot aqueous solution (enrichment by selective dissolution, transport, and precipitation, e.g., Heinrich and Candela, 2014), which may have had a magmatic, metamorphic, surface, or other sources (Robb, 2005). In other words for the

As-Bi	As-Bi-(Zn)-(Mo)	Zn+Mo
	Sc	Sc
Nb/Ta	Nb/Ta	Nb/Ta+Na
Li (Si)	Li (Si) ⇒Li (P) ⇒Li (B)	
P	P	P
B	B/F< 1⇒B/F>> 1	F
	Th/U<1	Th/U>1
	Zr	Zr-Ti
	REE	REE+Na
Be	Be	Be+Na
Sn-W?	Sn-W	Sn-W+Na

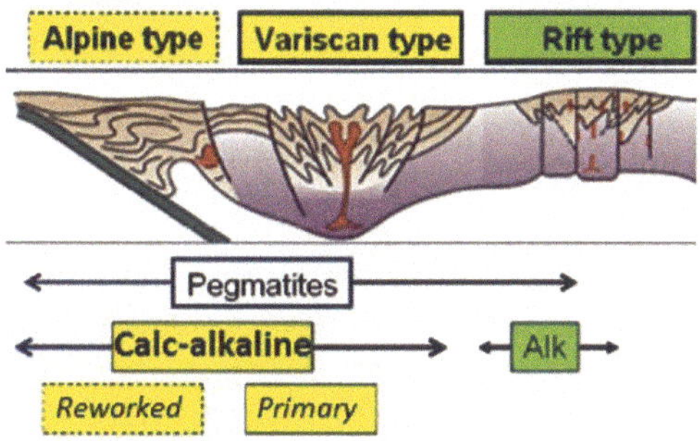

FIGURE 2.4 Rare element pegmatites and the geodynamic setting (CMS classification scheme—chemical qualifiers). The size and font of the letters are used to demarcate the significance of each element in the various settings. Bold-faced means widespread occurrence, set in brackets or added up with a quotation mark means minor potential as to the accumulation of a certain element or concentration processes uncertain. Source: *From Dill, H.G., 2015. Pegmatites and aplites: their genetic and applied ore geology. Ore Geol. Rev. 69, 417–561. Available from: https://doi.org/10.1016/j.oregeorev.2015.02.022.*

metals to be precipitated, they must have been selectively dissolved. How many metals are selectively dissolved is the function of several factors including the nature of metal–ligand complexes that can be formed (Robb, 2005), a process that can be well described by the Pearson's principle.

The Pearson's principle states that in a competitive situation, hard metals (acids or electron acceptors) will tend to complex with hard ligands (bases or electron donors), and soft metals with soft ligands. The Li^+, Na^+, K^+, Rb^+, Cs^+ for example, are hard metals, whereas Au^+, Ag^+, Cu^+ are the soft metals and Zn^{2+}, Pb^{2+}, Fe^{2+} are the divalent transition metals or borderline metals (Robb, 2005). Other hard metals described in Robb (2005) include Be^{2+}, Sr^{2+}, Ba^{2+}, Fe^{3+} and Ce^{4+}, Sn^{4+}, Mo^{4+}, W^{4+}, V^{4+}, Mn^4 and soft metals include Hg^{2+},

TABLE 2.3 Ligands occurring in hydrothermal ore solutions (Seward et al., 2014).

Hydroxide	OH^-
Halide ions	F^-, Cl^-, Br^-, I^-,
Sulfur species	HS, Sn, SnS^{2-}, $SO3^{2-}$, $S_2O_3^{2-}$, SO_4^{2-}
Ammonium (ammine)	NH3
Oxyanions	CO_3^{2-}, PO_4^{3-}, ASO_3^{3-}, SbO_3^{3-}, MoO_4^{2-}, WO_4^{2-}, SiO_4^{4-}
Thioanions	AsS_3^{3-}, SbS_3^{3-}, MoS_4^{2-}, WS_4^{2-}
Carboxylates	CH_3COO^- (acetate), $C_2H_5COO^-$ (propionate), $CH_2(COO)_2^{2-}$ (malonate), $(COO)_2^{2-}$ (oxalate)
Miscellaneous ligands of possible interest	HTe^-, Te_2^{2-}, CN^-, SCN^-

Cd^{2+}, Sn^{2+}, Pt^{2+} Pd^{2+} and Hg^{2+}, Cd^{2+}, Sn^{2+}, Pt^{2+}, Pd^{2+}. An example of hard ligand is $OH^- - F^- - NO_3^- - HCO_3^- CH_3COO^-$, whereas $HS^- - I^- - CN^- - H_2SS_2O_3^{2-}$ is a soft ligand.

The ligands of interest in hydrothermal systems are shown in Table 2.3. These metals and ligands are described in terms of their hard–soft breakdown and how that relationship is applicable to ore-forming processes. The metal–ligand relationship provided above is simply an approximation of what is most likely to happen in the function of variables such as oxidation state, pH, temperature, and fluid composition (Robb, 2005). In nature, metals may complex with any base, most suitable available, thus the metal–ligand complex to be formed will be quite different to those predicted on theoretical grounds (Robb, 2005).

Brugger et al. (2016) show that apart from metal–ligand complexes being available for ore deposition, the ore transport and precipitation of metals are largely controlled by coordination chemistry of hydrothermal systems as shown in the periodic table for hydrothermal geochemistry in Fig. 2.5.

However, in economic geology, what is important is not just the formation of the metal but the formation of a viable ore deposit. It requires the presence of high fluid/rock ratios, as well as efficient precipitation mechanisms to take metals out of solution and concentrate them in the host rock (e.g., Seward and Barnes, 1997). For example, solutions that deposit Au and Ag are typically in the range 1 ppb to 1 ppm, than those associated with Cu, Pb, and Zn ores. In the latter case, massive sulfide deposits are formed from fluids that typically carry 1–100 ppm of metal, whereas the fluids associated with the MVT and PDs are relatively enriched with concentrations in the range 100–1000 ppm.

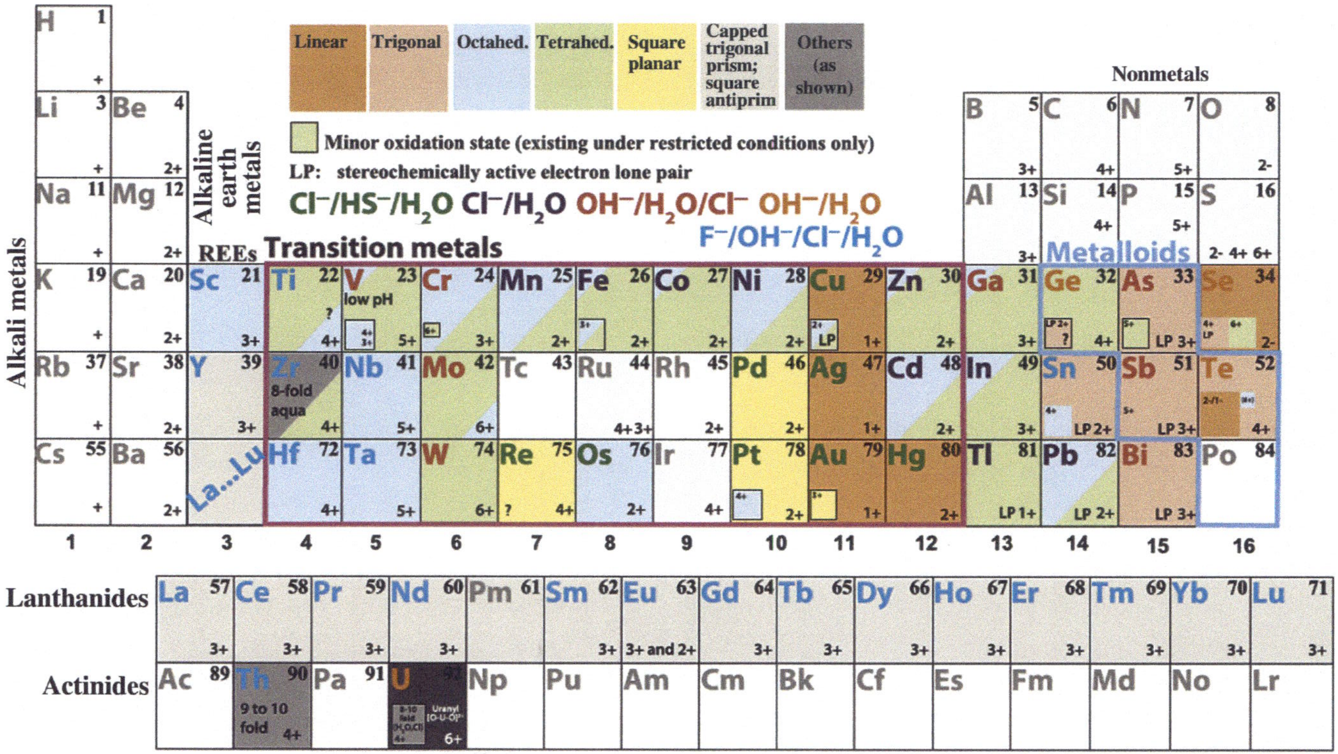

FIGURE 2.5 A periodic table for hydrothermal geochemistry showing coordination chemistry of metals in upper crustal geological fluids. Source: *From Brugger, J., Liu, W., Etschmann, B., Mei, Y., Sherman, D.M., Testemale, D., 2016. A review of the coordination chemistry of hydrothermal systems, or do coordination changes make ore deposits? Chem. Geol. 447, 219–253. Available from: https://doi.org/10.1016/j.chemgeo.2016.10.021.*

Skarn deposits (e.g., W, Cu, Au), PDs (e.g., "Cu, Mo porphyries") and vein deposits (e.g., Sn–W veins) are actually magmatic–hydrothermal deposits because they are caused by fluid saturation of H_2O-bearing magmas (Heinrich and Candela, 2014). Other examples of hydrothermal deposits include Epithermal veins and breccias that are due to continental geothermal systems (i.e., shallow magmatic fluids), Iron oxide hosted Cu–Au–U–REE ("IOCG") deposits caused by continental magmatic fluids interacting with evaporitic brines, volcanic-hosted massive sulfide deposits due to seawater convection through oceanic crust, orogenic ("mesothermal") gold–quartz veins due to metamorphic dehydration of deep magmatic fluids, sediment-hosted Cu ± Co deposits and stratiform Pb–Zn–Ag deposits ("sedex" = sedimentary–exhalative), caused by basin brines: connate or meteoric water + evaporites (Heinrich and Candela, 2014). Other examples of hydrothermal deposits due to basin brines are the MVT Pb–Zn + epigenetic replacement in carbonates ± sandstones. Uranium in sandstones ("roll-front U deposits") and vein—and "unconformity-related U deposits" are also hydrothermal deposits caused by deep infiltration of oxygenated surface waters.

1. Deposits due to lateral secretion. In this type of deposits, ore- and gangue-forming materials from the country rocks diffuse into faults and other structures (Moon, 2016).

Pyrometasomatic deposits are formed by both magmatic and hydrothermal processes, that is, magmatic–hydrothermal ore-forming processes. These deposits can be facilitated by some physical and chemical properties of water such as magmatic–hydrothermal fluids, water solubility in magmas, and first boiling and second boiling of granite-related magmatic–hydrothermal ore deposits. Minerals in this group include porphyry Cu, Mo, and W deposits, polymetallic skarn deposits, and epithermal Au–Ag–(Cu) deposits (Fig. 2.7). Generally, Magmatic–hydrothermal deposits are the sole producers of global Cu, Mo, Sn, W, In, and Re, and are a significant source of Au, Ag, Pb, Zn, and other minor and rare metals (Heinrich and Candela, 2014). Summarized geological context of PD and epithermal ore deposits are shown in Fig. 2.6.

Regions that experienced such kind of deposits include the magmatic–hydrothermal fluids associated with granite intrusions of the La Escondida porphyry Cu deposit in Chile, the MacTung W skarn deposit in Yukon–Canada, and magmatic–hydrothermal fluids in volcanic environments such as the high- and low-sulfidation epithermal gold deposits of Kyushu, Japan and Witwatersrand Basin gold deposit, in South Africa (Tucker et al., 2016).

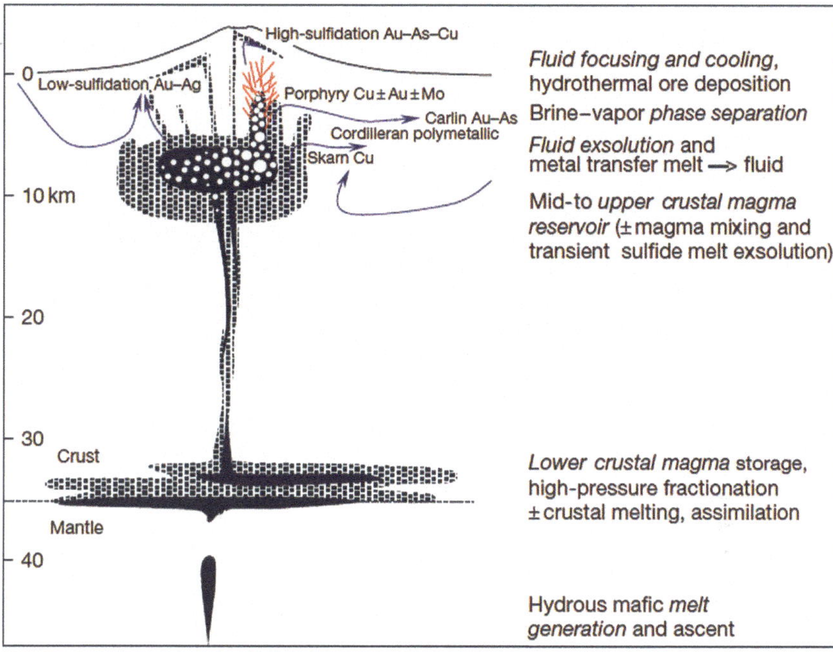

FIGURE 2.6 Schematic illustration of the geological components, ore types, and processes in crustal-scale magmatic–hydrothermal systems. Porphyry copper deposits form in a dense network of hydrofractures (red) around dikes and stocks above the roof of hydrous magma chambers, by precipitation of sulfides and gold from single phase or more commonly two-phase fluids of coexisting hypersaline liquid (brine) and vapor. Au-rich varieties predominate at shallow levels (1–3 km below surface) while porphyry deposits formed as deep as 8 km are generally Cu ± Mo dominated and Au-poor. Copper ± Zn ± Au skarns, Cordilleran vein, and replacement Cu–Zn–Pb–Ag ± Au deposits, and Carlin-type Au deposits rich in arsenious pyrite are hosted by carbonate-bearing sedimentary host rocks controlling pH by carbonate dissolution; they form at decreasing temperatures and increasing distance from magmatic intrusions. Carlin-type gold deposits are additionally characterized by abundant organic carbon in the host sediments. Their formation temperature and the low salinity of ore fluids with liquid-like density overlap with fluid properties in other epithermal precious metal deposits. High-sulfidation epithermal deposits are characterized by intense acid leaching and sulfate and clay alteration caused by hot low-density magmatic vapor, typically preceding the main stage of ore deposition by low-salinity aqueous liquid. Low-sulfidation epithermal deposits are generally more distal to intrusions and are characterized by near-neutral fluids causing feldspar–muscovite alteration, with a greater proportion of convecting meteoric water and a less obvious magmatic fluid input. Source: *From Heinrich, C.A., Candela, P.A., 2014. Fluids and ore formation in the Earth's crust. In: Treatise on Geochemistry, pp. 1–28.*

2.1.2.4 Deposits due to external surface processes

Ore deposits in this category can be due to mechanical accumulation, sedimentary precipitates, residual processes, secondary supergene environment, and volcanic exhalative (=sedimentary exhalative):

- Deposits due to mechanical accumulation: This is due to the concentration of heavy, durable minerals into placer deposits (Guilbert and Park, 1986). Examples of minerals deposited by this process are rutile–zircon sands, Tin placers, gold placers, industrial sands, gravels, and kaolin deposits (Moon, 2016).
- Deposits due to sedimentary precipitates: This is due to precipitation of particular elements in suitable sedimentary environments, with or without the intervention of biological organisms (Westphal et al., 2010). Minerals deposited by this process include banded iron formation of Precambrian shields, Mn deposits, evaporite deposits, and phosphate deposits (e.g., Moon, 2016).
- Deposits due to residual processes: These are due to soluble elements in the remaining material are leached from rocks and get deposited somewhere where they form deposits (Guilbert and Park, 1986). Examples of minerals deposited by this process include gossan deposits of Ni laterites, bauxites, and kaolin deposits (Moon, 2016).
- Deposits due to secondary supergene environment: These kinds of deposits are made possible by leaching of valuable elements from the upper parts of mineral deposits and their precipitation at depth to produce higher concentrations. Examples of minerals in this subcategory include many gold and silver bonanzas and the upper parts of a number of porphyry copper deposits (e.g., Robb, 2005; Moon et al., 2006).
- Deposits due to volcanic exhalative (=sedimentary exhalative): These kinds of deposits are due to exhalations of hydrothermal solutions at the surface, usually under marine conditions and generally producing stratiform ore bodies (Robb, 2005). Examples are base metal deposits of Megagan, German; Kuroko deposits of Japan; and Solfatara deposits (kaolin + alunite).

2.1.3 The role of global tectonics and geological time in the formation of ore deposits

Global tectonics is the second major factor in the formation of distinct ore deposit types and a key to understanding their spatial and secular distribution (Groves et al., 2005; Holland, 2005) as most hydrothermal environments where ores deposits are formed are commonly found in many active, continental, and oceanic geothermal fields along plate tectonics margins (Barnes, 2015). Thus chromium deposits form by physical accumulation of what is usually an accessory mineral (chromite, $FeCr_2O_4$) from large reservoirs of mafic magma. Such magma chambers either form as intraplate layered intrusions (possibly related to mantle plumes) or at spreading axes in the oceanic domain (Stowe, 1994).

The world porphyry Cu–Au belt, for example, is restricted within recent converging/subduction zones. Worldwide, Cu–Au mineralization seems to be restricted within ridge subduction zones that are characterized by adakitic rocks (Sun et al., 2010: 17). Porphyry Cu mineralization seems to be restricted in arc-related structures within the Phanerozoic. This coincides well with the ring of fire. The reason why plate boundaries are the locus of mineralization is that the energy released along boundaries, results in volcanic and hydrothermal activity that creates the right conditions for many minerals to be concentrated. Plate boundaries are also a way for deep-seated rocks from the mantle to find a way to the near-surface because these boundaries, as earlier on hinted, are the weak zones. These rocks become the source of many of our largest mineral deposits. Geologists therefore have to know how the plates move and predict where they could have been in the past—by having a clear understanding of the location of ancient plate boundaries, it is easier to predict where regional mineralization, and hence ore deposits may occur today (Arribas and Mizuta, 2018).

The ore deposit types with global tectonic settings are shown in Fig. 2.7. It should be noted however that global tectonics does not answer all questions related to ore deposits. In other words, location of mineralization on the crust is not necessarily the function of plate tectonics; in fact, some deposit types appear to require the absence of plate tectonics inasmuch as they form tectonically undisturbed regimes (Sangster, 1980). Examples include the MVT-type Pb–Zn deposits, sandstone U deposits, sedimentary barites, phosphates, and superior-type iron formations; they are not easily explained by global tectonics.

In summary, three groups of factors—transport medium, global tectonics, and the earth's changing climate and atmosphere—allow a first-order explanation for the range of major ore deposit types. Common to all ore-forming processes is the fact that selective element enrichment is an entropy-reducing process that requires a net input of energy. For a deeper understanding of the factors controlling ore formation, we therefore need to focus on the energy sources—that is, the physical and chemical driving forces—contributing to processes separating elements from each other on a large scale and with high efficiency (Heinrich and Candela, 2014) (Fig. 2.8).

2.1.4 Source of metals in ore deposits: principles

Understanding the source of metals of ore deposits is paramount as such knowledge can assist in predicting or locating similar mineralized parts of the earth's crust. Temperature, pressure, oxygen fugacity are important variables in understanding deposit genesis (Kotzer and

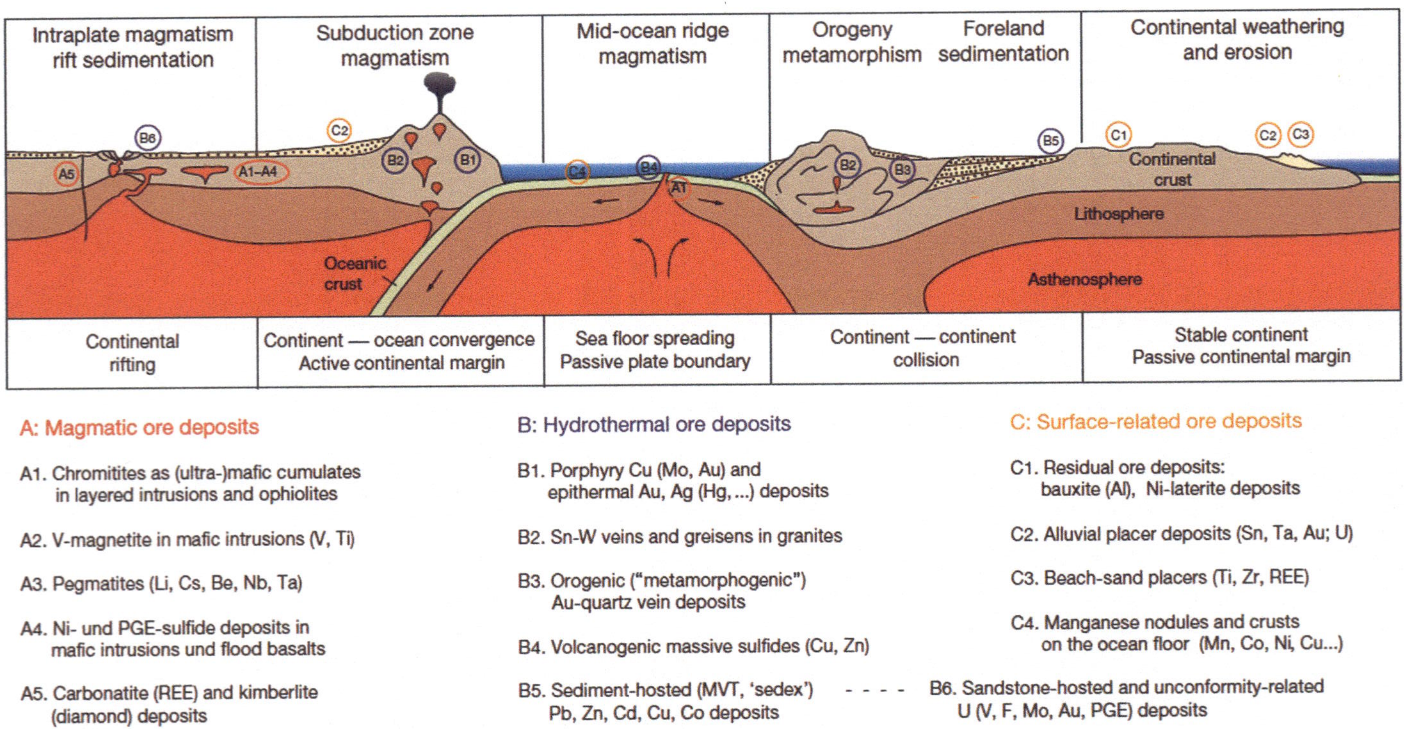

FIGURE 2.7 Schematic illustration of the recurrent association of ore deposit types with global tectonic settings, including active and passive continental margins, oceanic spreading centers including back-arc basins, as well as land surfaces, sedimentary basins, and hot spots in the interior of lithospheric plates. Labels A, B, C refer to major metal-transporting media and ore deposit types in Table 2.3. Source: *From Heinrich, C.A., Candela, P.A., 2014. Fluids and ore formation in the Earth's crust. In: Treatise on Geochemistry, pp. 1–28.*

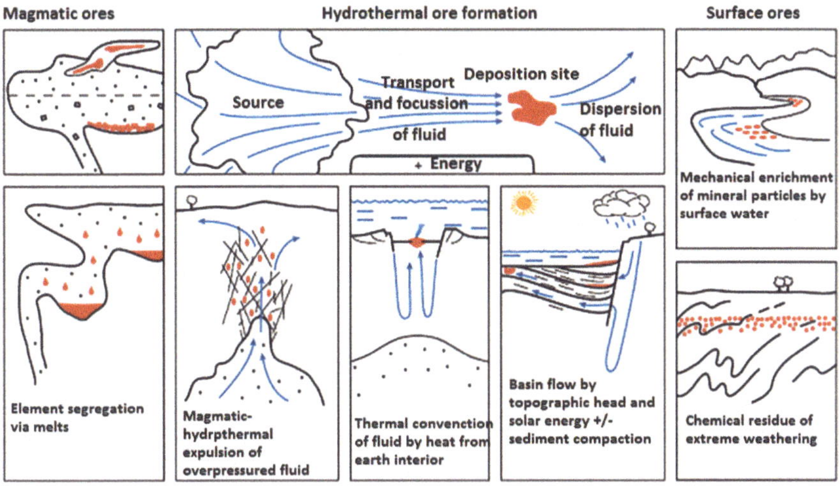

FIGURE 2.8 Schematic illustration of three principal processes of selective element enrichment that can produce mineral deposits in the earth's crust, grouped by the main agent of material transport: magmatic melts, hydrothermal fluids, or surface waters (Kesler, 1994). Source: *From Heinrich, C.A., Candela, P.A., 2014. Fluids and ore formation in the Earth's crust. In: Treatise on Geochemistry, pp. 1–28.*

Kyser, 1995). Isotopic studies (Kotzer and Kyser, 1995; Legros et al., 2019), fluid inclusions (Richard et al., 2013; Brugger et al., 2016; Kyser, 2016; Xiong et al., 2018), and trace element geochemistry are important aspects to be considered in studying deposit genesis (e.g., Robb, 2005; Mungall, 2014). The principles underlying these studies, particularly those on geochemical techniques, are briefly discussed in this subchapter.

2.1.4.1 *Pressure, temperature, chemical composition, and oxygen fugacity*

Pressure, temperature, chemical composition, and oxygen fugacity (*f*O2) are critical parameters useful in understanding ore deposit characteristics and processes that lead to ore generation and they can assist in identifying correct environment (s) for exploring potential minerals (Legros et al., 2019). In the context of chemical composition for example, most highly evolved alkaline intrusives do not host uranium deposits because they do not evolve to concentrate uranium in their differentiates (Kotzer and Kyser, 1995; Richard et al., 2013; Xiong, et al., 2018).

Regarding oxygen fugacity, Kress and Carmichael (1991) show that the Fe_2O_3/FeO ratio of melts is linked to its oxygen fugacity and can be estimated by the empirical relationship:

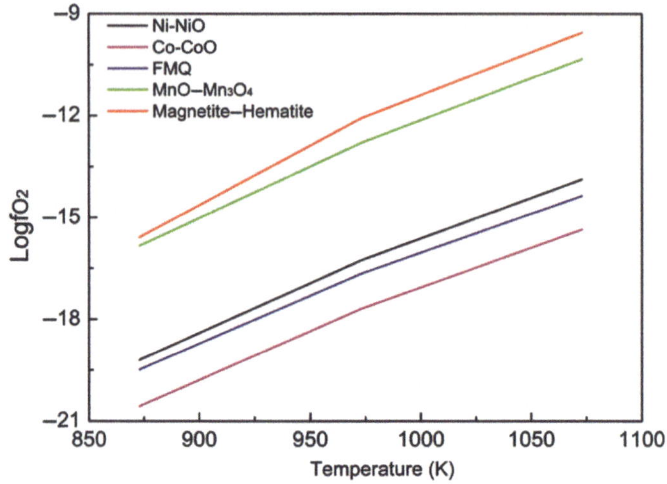

FIGURE 2.9 Fugacity–temperature diagram. Log *f*O2, versus T at 1 bar P for common buffer assemblages. MH, NiNiO, FMQ, WM, IW, and QIF are respectively magnetite–hematite, nickel–nickel oxide, fayalite–magnetite–quartz, wüstite–magnetite, iron–wüstite and quartz–iron–fayalite (e.g., Chou, 1978; Lindsley, 1991; Frost et al., 1988). Source: *From Sun, W., Huang, R., Li, H., Hu, Y., Zhang, C., Sun, S., et al., 2015. Porphyry deposits and oxidized magmas: a review. Ore Geol. Rev. 65, 97–131.*

$$\log\left(fO2\right) = \log\left\{\exp\left[\frac{\left(\frac{\ln(Fe_2O_3)}{FeO}melt - \frac{b}{T} - c - \sum_i^n diXi\right)}{a}\right]\right\} \tag{2.2}$$

where $a = 0.207$; $b = 12{,}980$; $c = -6.115$. Temperature T is in Kelvin and Xi is between 0 and 1 calculated from oxides as wt.% (Liao et al., 2016).

Further, in silicate magma and in ideal cases, oxygen fugacity is equal to its partial pressure (e.g., Evans et al., 2017) and positively correlated with temperature, T (Lindsley, 1991). Fig. 2.9 illustrates how oxygen fugacity can be related to mineral assemblage buffers such as fayalite–magnetite–quartz (FMQ), nickel–nickel oxide, that is, Ni–NiO, wüstite–magnetite (WM), iron-wüstite (IW); quartz–iron–fayalite (QIF), and hematite–magnetite (e.g., Chou, 1978; Lindsley, 1991; Frost et al., 1988; Ablay et al., 1998).

Mineral assemblage buffers are assemblages of minerals or compounds that constrain oxygen fugacities as a function of temperature (Lindsley, 1991). If a rock contains pure minerals that constitute a redox buffer (e.g., magnetite–hematite, nickel–nickel oxide, FMQ, WM, IW, QIF), then the oxygen fugacity of equilibration is defined by one of the curves in Fig. 2.9 (Chou, 1978; Lindsley, 1991; Frost et al., 1988). For other rocks with suitable minerals, oxygen fugacities can be calculated, and the redox conditions that are obtained from calculation be compared by the fugacity–temperature diagram in Fig. 2.9 (e.g., Lindsley, 1991).

Oxygen fugacity strongly influences the crystallization sequences and composition of crystallizing minerals (Liao et al., 2016). Preservation of diamonds during transport from the mantle to the surface is estimated by assessing the oxygen fugacity of the magma (Moon et al., 2006). Highly oxidized intrusions and economic ore deposits are commonly associated with relatively higher oxygen fugacity values (Trail et al., 2012; Zhou et al., 2018). For example, the oxygen fugacity value of ΔFMQ +2 averaging between +4.50 and +7.12 is associated with medium-sized to super large-sized porphyry Mo deposits in the East Xing'an–Mongolian Orogenic Belt and below which the deposits are small (Zhou et al., 2018). It should be emphasized here that not all mineral deposits are associated with increase in oxygen fugacity as some minerals would be deposited when oxygen fugacity decreases. This is because oxygen fugacity is not the only control of mineral deposition (e.g., Chou, 1978). Other factors that control mineral deposition are pressure, temperature pH, Eh, sulfur fugacity, etc. (Fig. 2.10). For example, gold mostly occurs as $AuCl_2^-$ in a system with temperature higher than 400°C (Gammons and Williams-Jones, 1997; Zhu et al., 2011), but as temperature decreases (the primary mechanism), gold is deposited (Zhu et al., 2011). At lower temperatures, $Au(HS)_2^-$ is the dominant

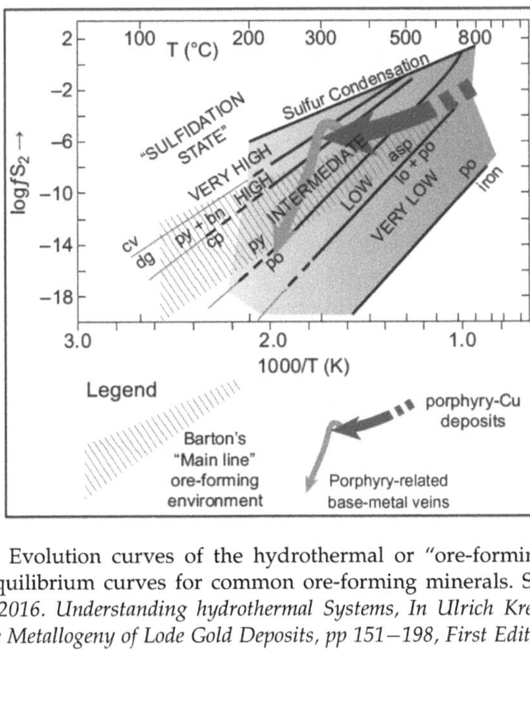

FIGURE 2.10 Evolution curves of the hydrothermal or "ore-forming" fluids. Shown are sulfidation equilibrium curves for common ore-forming minerals. Source: *Kretschmar, U., McBride, D., 2016. Understanding hydrothermal Systems, In Ulrich Kretschmar and Derek McBride (eds), The Metallogeny of Lode Gold Deposits, pp 151–198, First Edition Elsevier, 345p.*

phase, and the maximum solubility exists in the vicinity of the $H_2S - HS^- - SO_4^{2-}$ equilibrium point and with the decline of oxygen fugacity, the Au–S complex breaks down leading to gold precipitation (Cooke and Simmons, 2000; Robb, 2005; Zhu et al., 2011):

$$2Au(HS)_2^- + H_2O \rightarrow 2Au + 4HS^- + 2H^+ + 0.5O_2$$

$$2Au(HS)_2^- + 8H_2O \rightarrow 2Au + 4SO_4^{2-} + 4H^+ + 8H_2$$

$$FeCO_3 + Au(HS)_2^- = FeS_2 \text{ pyrite} + CO_2 + H_2O + Au \text{ (Zhu et al., 2011)}$$

Iron sulfides monitor the sulfur fugacity just like how Fe–Ti oxides monitor oxygen fugacity in rocks (Frost and Frost, 2014). With increasing temperature, or at low sulfur fugacities, pyrite breaks down to pyrrhotite by the reaction:

$$(1 - x)\,FeS_2 = Fe(1 - x)S + \left(\frac{1}{2 - x}\right) S_2 \tag{2.3}$$

This relationship indicates that, whereas pyrite may be stable in some relatively low-temperature rocks such as granites, pyrrhotite is the only sulfide found in high-temperature rocks such as basalt and gabbro.

The concentration of gold in potentially ore-forming solutions is a function of the distribution of sulfur species. For temperatures of 200°C–400°C, pressures of 200 MPa and near-neutral pH, $Au(HS)_2$—is the dominant gold–hydrosulfide complex (Shvarov and Bastrakov, 1999; Phillips and Evans, 2004). Thus sulfur is one of the key elements trapped in fluids during movements of hydrothermal fluids as a function of temperature, pressure, and pH (e.g., Phillips and Evans, 2004; Frost and Frost, 2014). Trisulfur ion, (S^{-3}), recently discovered in laboratory experiments (Pokrovski and Dubessy, 2015), may account for up to 10% of total dissolved sulfur (S_{tot}) at 300°C–500°C in fluids from arc-related magmatic–hydrothermal systems, and more than 50% S_{tot} at 600°C–700°C in S-rich fluids produced via prograde metamorphism of pyrite-bearing rocks (Steudel and Steudel, 2013; Pokrovski and Dubessy, 2015). In addition, the trisulfur ion may favor the mobility of sulfur itself and associated metals (Au, Cu, Pt, and Mo) in geological fluids over a large range of depth and provide the source of these elements for orogenic Au and porphyry–epithermal Cu–Au–Mo deposits (Chivers, 1974; Steudel and Steudel, 2013; Pokrovski and Dubessy, 2015).

2.1.4.2 *Fluid inclusions*

Fluid inclusions are microscopic pockets of gas or liquids trapped within minerals (Craig and Vaughan, 1981; Sun et al., 2015; Corral et al., 2017; Fig. 2.11). The pockets may contain information on the original

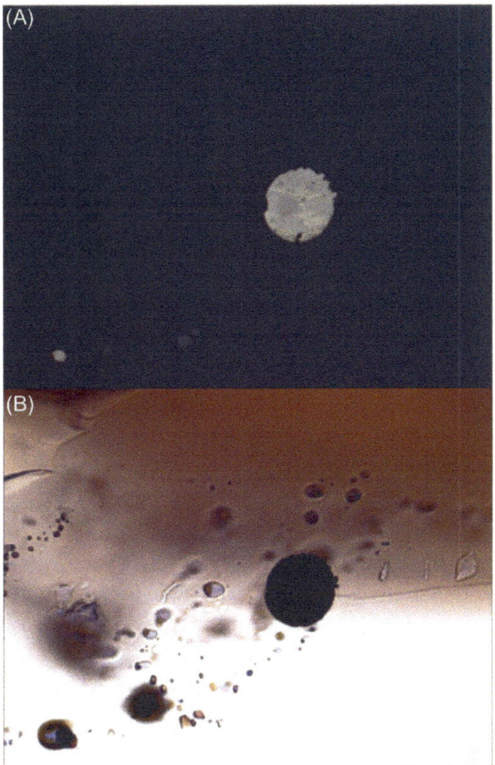

FIGURE 2.11 Images of sulfide inclusion in an olivine phenocryst of Manus glass under (A) reflected light and (B) and transparent light. Only one big sulfide globule is clearly identified under reflected light, which seemingly indicates sulfide saturation. This grain is actually associated with fluid inclusions as shown under transparent light. The unique composition of sulfides in Manus glasses (low Ni, Pt, etc.) may be plausibly interpreted by sulfides crystallized from magmatic fluids (Sun et al., 2015).

physical and chemical conditions at which the source rock was formed (e.g., Viter et al., 2013) and how the transportation–deposition mechanism of ore-forming fluids (in hydrothermal deposits) acted (Bodnar et al., 2014). The fluid inclusions can also constrain geochemical models used to investigate geochemical processes (Bertelli et al., 2009). Fluid inclusions, when trapped within crystals during initial growth from solution or during total recrystallization, are called primary inclusions or if recrystallization of the fluid inclusions is localized along fractures at some later time, they are referred to as secondary inclusions (Craig and Vaughan, 1981). In fluid inclusion studies therefore the fundamental task is to study the nature of fluids included or trapped by the minerals during alteration and mineralization processes (Trail et al., 2009, 2012; Xiong, et al., 2018).

2.1.4.3 Radiogenic and nonradiogenic (stable) isotopes

Isotopes are used in bedrock geology, archeology, forensics, food and beverage industry, and environmental sciences. The focus of this book and in particular this subchapter is on the application of isotopes in the field of geology, more especially in understanding the geodynamics of rocks, the ore deposit sources, and age of rocks and their associated mineral deposits.

The U–Pb (from sphalerite) and Re–Os (from molybdenite) are examples of sets of radiogenic elements that can be used to understand the geodynamic setting of some mineral deposits (e.g., Robb, 2005; Dai et al., 2009) and their age when properly employed. For instance, Dai et al. (2009) use the stable isotopes of sulfur ($\delta^{34}S$: 3.3‰–8.0‰) to predict that a mixed crust–mantle system was responsible for the metal ore source in the Mo (Fe) deposit Xiaojiayingzi, China. Furthermore, using the same data, they propose an early stage of subduction of the Paleo-Pacific plate beneath the Eurasian Block as related to the geodynamic setting of that deposit. Casa et al. (2003) employ a similar approach in determining the source of gold in Central Italy. They propose that the origin of gold in Ponte San Pietro deposit may be related to leaching of metapelites/phyllites derived from pelitic rocks containing some amount of sulfur-bearing organic matter. The Co, Ni, Mn, as well as the $\delta^{34}S$ data ($\delta^{34}S$ values of up to 7.0‰), suggest a different origin for the phyllite-hosted marcasite, the Ponte San Pietro quartz-vein-hosted pyrite, and for gold occurrences (see Table 2.4).

TABLE 2.4 Distribution of elements, element ratios, and $\delta^{34}S$‰ in marcasite within phyllites of the Verrucano formation (T–D samples) and in vein-hosted pyrite from Ponte San Pietro deposit, Central Italy.

Sample	Au	Co	Ni	Mn	As	Co/Ni	$\delta^{34}S$‰
T-D1	424	41	56	390	236	0.73	+ 2.5
T-D2	54	43	62	400	182	0.69	+ 7.0
T-D3	338	45	60	300	187	0.75	nd
T-E3	7400	Nd	nd	nd	nd	nd	+ 0.8
T-E4	1400	26	26	10	7500	1.0	+ 1.9
T-E5	2000	32	22	8	8500	1.45	nd
T-E6	17,000	47	29	13	71,000	1.62	nd
T-E8	21,000	28	18	16	8100	1.55	nd

Element values are given in ppm, except for Au in ppb, *nd*, not determined.

From Casa, G.D., Manni, A., Saviano, G., Violo, G., 2003. Gold occurrence in Central Italy—the Ponte San Pietro mineralization. Ore Geol. Rev. 23, 99–105. Available from: https://doi.org/10.1016/S0169-1368(03)00017-9.

The expression $\delta^{34}S = x‰$ means that ^{34}S is in parts per thousand and expressed relative to the most abundant sulfur isotope $\delta^{32}S$ (e.g., Ridley, 2013), that is:

$$\delta^{34}S = \left[\frac{\left(\frac{^{34}S}{^{32}S}\right)\text{sample} - \left(\frac{^{34}S}{^{32}S}\right)\text{standard}}{\left(\frac{^{34}S}{^{32}S}\right)\text{standard}} \right] \times 1000 \qquad (2.4)$$

It is worth noting that in nature, the stable sulfur isotopes ^{32}S, ^{33}S, ^{34}S, and ^{36}S are approximately 95%, 0.75%, 4.2%, and 0.02%, respectively. The standard for sulfur isotope ratio stated in Eq. (3.3) is of iron meteorite and mantle-derived sulfur and has $\delta^{34}S$ values within a narrow range of 0% ± 1% (Ridley, 2013).

Isotopes of H and O can be used to tell about the origin of hydrothermal fluids as to whether or not they are from seawater, meteoric water, magmatic water, connate water, metamorphic water, or from mixed sources in the earth's crust that could have been responsible for the mineralization in question (e.g., Robb, 2005; Corral et al., 2017). The plot of hydrogen (δD permil) versus oxygen ($\delta^{18}O$ permil) isotopic ratios for various water types have been shown in Fig. 2.12. Note from that plot the linear meteoric water line ($\delta D = 8 \times \delta^{18}O + 10$) as well as the standard mean seawater values, standard mean ocean water (SMOW; δD permil = 0, $\delta^{18}O$ permil = 0). Thus one has to analyze from the ore samples, the values of δD and $\delta^{18}O$ and see where they plot in Fig. 2.12 for establishing the source of hydrothermal fluids. As it is always not easy to date the ore itself due to the absence of datable material, dating is done by using nonore materials associated with the ore. The plot of $^{143}Nd/^{144}Nd$ versus $^{87}Sr/^{86}Sr$ has been used by White and Patchett (1984) to characterize island arc volcanic materials of the South Sandwich, Lesser Antilles, Aleutians, Marianas, Philippines, Taiwan, Banda, and Sunda arcs (Fig. 2.13).

Further, the plot of $\delta^{18}O$ (‰) versus $\delta^{34}S$ (‰) for the Cerro Quema Au–Cu deposit (Azuero Peninsula, Panama) show that the fluid in the magmatic–hydrothermal system was sulfide dominant ($XH_2S = 0.69$) and the sulfur was of magmatic origin ($\delta^{34}S_{\Sigma S} = -0.5‰$) but that variable $\delta^{18}O$ at constant $\delta^{34}S$ of alunite and barite, and $\delta^{18}O$ values of fluids being in equilibrium with vuggy quartz (−2.3 to 3.1‰) suggest dilution of magmatic fluid by meteoric waters (e.g., groundwater) during alteration/mineralization (Corral et al., 2017). The $\delta^{13}C$ isotopic compositions for the DRC diamonds show values from −12.5‰ to −1.9‰ implying near mantle-like values (Kosman et al., 2016).

Deuterium–oxygen–carbon (D–O–C) isotopic systematics have been used by some scholars such as Aliyari et al. (2009) to study the hydrothermal deposits.

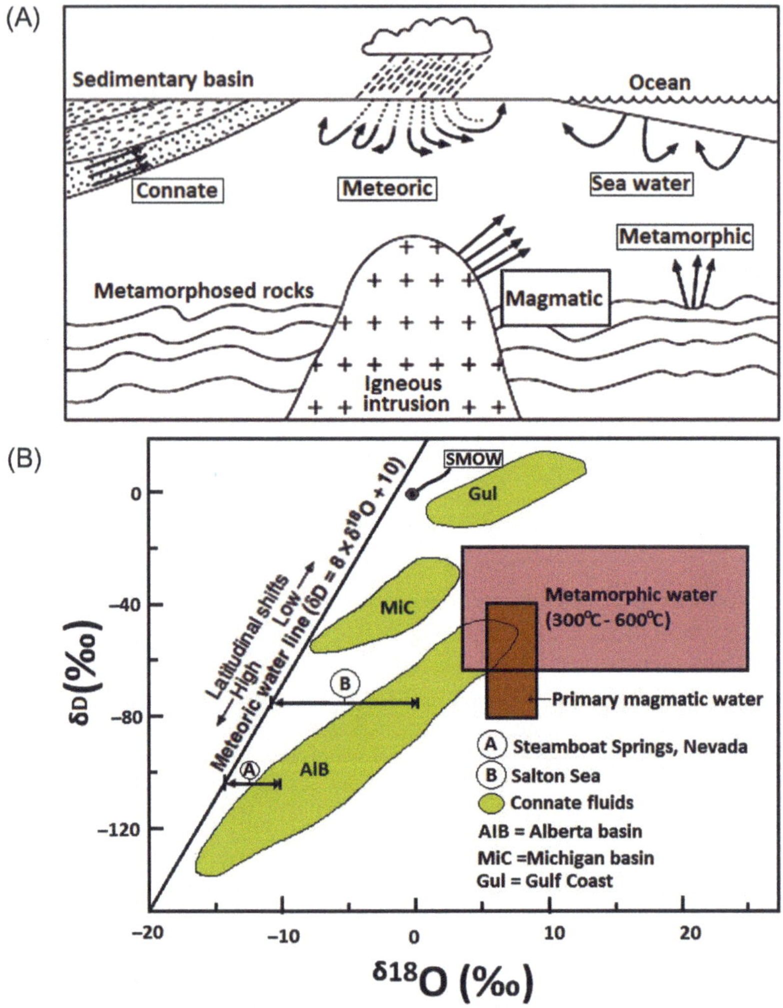

FIGURE 2.12 (A) The major types of liquid water that exist at or near the earth's surface. (B) Plot of hydrogen (δD permil) and oxygen (δ^{18}O permil) isotopic ratios for various water types. SMOW is defined to be zero for both δD and δ^{18}O. Some ore-forming environments are clearly related to mixed fluid reservoirs; in the cases shown (see A and B) mixing of meteoric and connate fluids has taken place. Source: *After Taylor, H.P., 1997. Oxygen and hydrogen isotope relationships in hydrothermal mineral deposits. In: Barnes, H.L. (Ed.), Geochemistry of Hydrothermal Ore Deposits, third ed. John Wiley & Sons, New York, pp. 229–302. From Kretschmar, U., McBride, D., 2016. Understanding hydrothermal Systems, In Ulrich Kretschmar and Derek McBride (eds), The Metallogeny of Lode Gold Deposits, pp 151–198, First Edition Elsevier, 345p.*

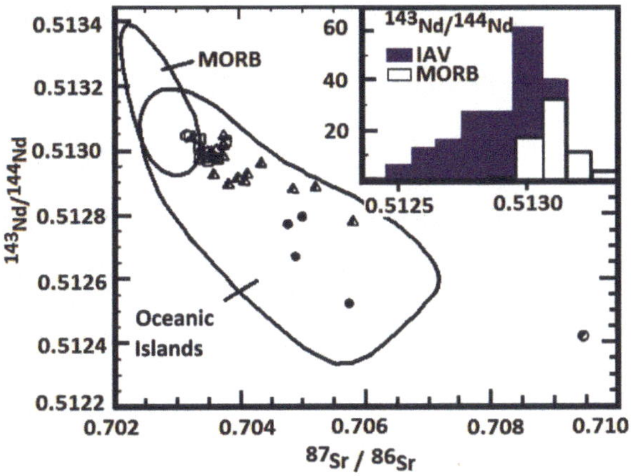

FIGURE 2.13 Nd–Sr isotopic data for Island Arc volcanic. Symbols are O = Aleutians, □ = New Britain, Δ = Marianas, ◪ = Izu, ● = Sunda, ▲ = Banda, and ◐ = Lesser Antilles. Inset refers to distinction in $^{143}Nd/^{144}Nd$ between mid-oceanic ridge basalts (MORB) and island arc volcanics (IAV) for all available analyses. Source: *From White, W.M., Patchett, J., 1984. HfNdSr isotopes and incompatible element abundances in island arcs: implications for magma origins and crust-mantle evolution. Earth Planet. Sci. Lett. 67 (2), 167–185. Available from: https://doi.org/10.1016/0012-821x(84)90112-2.*

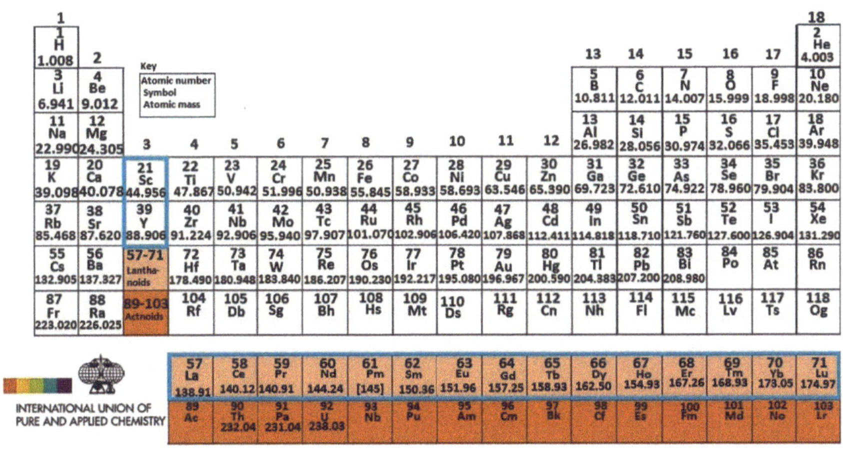

FIGURE 2.14 Periodic table of elements with rare earth elements, REE (LREE and HREE) shown in blue boxes. Source: *After Gaffney, J.S., Marley, N.A., 2018. Periodic table. In: General Chemistry for Engineers, 623. Available from: https://doi.org/10.1016/b978-0-12-810425-5.09983-5.*

2.1.4.4 Rare earth elements

2.1.4.4.1 Geochemistry and classification

REEs as indicated by blue boxes in Fig. 2.14 are a set of chemical elements in the periodic table, specifically the 15 lanthanides plus scandium

(Sc) and ythrium (Y), Sc, and Y are considered REE since they tend to occur in the same ore deposits as the lanthanides and exhibit similar chemical properties; their ability to readily discharge and accept electrons, make them indispensable and nonreplaceable in many electronic, optical, magnetic, and catalytic applications (e.g., Gaffney and Marley, 2018). Names of elements along with their symbols are shown in Table 2.5.

REEs are classified into two subgroups: LREEs comprising the first six elements of the Lanthanite series La, Ce, Pr, Nd, Pm, Sm (atomic numbers 57–62); and the heavy rare earth elements (HREE), comprising nine elements Eu, Gd, Tb, Dy, Ho, Er, Tm, Yb, Lu (with atomic numbers 63–71), and the 10th element ythrium despite its low atomic weight (Fig. 2.14), Y is classified with the HREE because its properties are closer to those of the HREE subgroup than to LREE (www.tasmanmetals.com/s/RareEarth.asp).

Despite their name, the REEs are in fact not especially rare. Each one of the REE is more common in the earth's crust than the elements Ag, Au or Pt, while Ce, Y, Nd, and La are more common than Pb (Fig. 2.14). Thulium and Lu are the least abundant REEs with crustal abundance of approximately 0.5 ppm.

The REEs, being capable of readily accepting and releasing electrons, are never found as free metals in the earth's crust and all their naturally occurring minerals consist of mixtures of various REEs with other metals and nonmetals. For example, garnet accommodates the HREE more than the LREE, and orthopyroxene and hornblende do so to a lesser degree. Sphene and plagioclase accommodate more LREE, and Eu^{2+} is strongly partitioned into plagioclase (e.g., Pun et al., 1997). Plotting concentration of the REE (normalized to mantle values) as the ordinate (*y*-axis) against increasing atomic number (as their degree of compatibility increases from left to right across the diagram), we can be able to characterize the genetic history of various rocks and ore deposits.

REE behave as incompatible elements, that is, they prefer magma over crystalline phases during crystal–liquid separation—be it partial melting or crystallization. As per Sen (2014), Sm and Nd behave as incompatible elements but Nd is slightly more incompatible than Sm. ^{143}Nd is a daughter isotope produced by α-decay of the radioactive isotope ^{147}Sm. Similar to the Sr system, Nd isotopic ratios are expressed in terms of $^{147}Sm/^{144}Nd$ (parent) and $^{143}Nd/^{144}Nd$ (daughter). Partial melting results in a greater Sm/Nd ratio of the residuum because Nd is more incompatible than Sm. It follows that as time progresses rocks crystallizing from such partial melts will evolve to lower $^{143}Nd/^{144}Nd$ than the residue because the residue will have a greater amount of $^{147}Sm/^{144}Nd$ ratio acquired during the partial melting event. Because of the differential change in Sm/Nd relative to Rb/Sr during partial

TABLE 2.5 Symbols of geochemical elements along with their names.

Atomic number	Element symbol	Name	Atomic number	Element symbol	Name
1	H	Hydrogen	60	Nd	Neodymium
2	He	Helium	61	Pm	Promethium
3	Li	Lithium	62	Sm	Samarium
4	Be	Beryllium	63	Eu	Europium
5	B	Boron	64	Gd	Gadolinium
6	C	Carbon	65	Tb	Terbium
7	N	Nitrogen	66	Dy	Dysprosium
8	O	Oxygen	67	Ho	Holmium
9	F	Fluorine	68	Er	Erbium
10	Ne	Neon	69	Tm	Thulium
11	Na	Sodium	70	Yb	Ytterbium
12	Mg	Magnesium	71	Lu	Lutetium
13	Al	Aluminum	72	Hf	Hafnium
14	Si	Silicon	73	Ta	Tantalum
15	P	Phosphorous	74	W	Tungsten
16	S	Sulfur	75	Re	Rhenium
17	Cl	Chlorine	76	Os	Osmium
18	Ar	Argon	77	Ir	Iridium
19	K	Potassium	78	Pt	Platinum
20	Ca	Calcium	79	Au	Gold
21	Sc	Scandium	80	Hg	Mercury
22	Ti	Titanium	81	Tl	Thallium
23	V	Vanadium	82	Pb	Lead
24	Cr	Chromium	83	Bi	Bismuth
25	Mn	Manganese	84	Po	Polonium
26	Fe	Iron	85	At	Astatine
27	Co	Cobalt	86	Rn	Radon
28	Ni	Nickel	87	Fr	Francium
29	Cu	Copper	88	Ra	Radium
30	Zn	Zinc	89	Ac	Actinium

(Continued)

TABLE 2.5 (Continued)

Atomic number	Element symbol	Name	Atomic number	Element symbol	Name
31	Ga	Gallium	90	Th	Thorium
32	Ge	Germanium	91	Pa	Protactinium
33	As	Arsenic	92	U	Uranium
34	Se	Selenium	93	Np	Neptunium
35	Br	Bromine	94	Pu	Plutonium
36	Kr	Krypton	95	Am	Americium
37	Rb	Rubidium	96	Cm	Curium
38	Sr	Strontium	97	Bk	Berkelium
39	Y	Yttrium	98	Cf	Californium
40	Zr	Zirconium	99	Es	Einsteinium
41	Nb	Niobium	100	Fm	Fermium
42	Mo	Molybdenum	101	Md	Mendelevium
43	Tc	Technetium	102	No	Nebelium
44	Ru	Ruthenium	103	Lr	Lawrencium
45	Rh	Rhodium	104	Rf	Rutherfordium
46	Pd	Palladium	105	Db	Dubnium
47	Ag	Silver	106	Sg	Seaborgium
48	Cd	Cadmium	107	Bh	Bohrium
49	In	Indium	108	Hs	Hassium
50	Sn	Tin	109	Mt	Meitnerium
51	Sb	Antimony	110	Ds	Darmstadtium
52	Te	Tellurium	111	Rg	Roentgenium
53	I	Iodine	112	Cn	Copernicium
54	Xe	Xenon	113	Nh	Nihonium
55	Cs	Cesium	114	Fl	Flerovium
56	Ba	Barium	115	Mc	Moscovium
57	La	Lanthanum	116	Lv	Livermorium
58	Ce	Cerium	117	Ts	Tennessine
59	Pr	Praseodymium	118	Og	Oganesson

Extracted from Fig. 2.14.

melting, the sign of the *epsilon* values for the Sm–Nd versus Rb–Sr systems means the opposite: positive $^{\varepsilon}$Nd value means depleted, whereas positive $^{\varepsilon}$Sr value means enriched composition (e.g., Sen, 2014; Kazemi et al., 2018).

Rocks from the primitive mantle have $^{\varepsilon}Nd = 0$ and $^{\varepsilon}Sr = 0$, whereas enriched igneous rocks have $^{\varepsilon}Nd < -2$ and $^{87}Sr/^{86}Sr > 0.705$ and depleted igneous rocks have $^{\varepsilon}$Nd of +6 to +10 and $^{87}Sr/^{86}Sr$ of 0.7025–0.703.

For kimberlites, although both Group 1 and Group 2 Kimberlites contain both compatible and incompatible elements, they can be distinguished from each other in the fact that Group 2 is enriched in LREE and depleted in Cr and Nb relative to Group 1 Kimberlites. The former is also H_2O rich, enriched in SiO_2, K_2O, Pb, Rb, and Ba (Sen, 2014).

References

Ablay, G.J., Carroll, M.R., Palmer, M.R., Marti, J., Sparks, R.S.J., 1998. Basanite–phonolite lineages of the Teide–Pico Viejo volcanic complex, Tenerife, Canary Islands. J. Petrol. 39 (5), 905–936.

Aliyari, F., Rastad, E., Mohajjel, M., Arehart, G.B., 2009. Geology and geochemistry of D–O–C isotope systematics of the Qolqoleh gold deposit, Northwestern Iran: implications for ore genesis. Ore Geol. Rev. 36 (4), 306–314.

Arribas, A., Mizuta, T., 2018. Potential for porphyry copper deposits in Northern Tohoku. Resource Geology, John Wiley & Sons. 68 (2), 144–163.

Bailey, R.A., Clark, H.M., Ferris, J.P., Krause, S., Strong, R.L., 2002. The earth's crust. In: Beiley, Ronald A. (Ed.), Chemistry of the Environment, second edition pp. 443–482, Academic Press, 835p.

Barnes, H.L., 2015. Hydrothermal processes. Geochem. Perspect. 4 (1), 1–93.

Bateman, A.M., 1950. Economic Mineral Deposits, second ed. John Wiley and Sons Inc, New York and London.

Begg, G.C., Hronsky, J.M.A., Griffin, W.L., O'Reilly, S.Y., 2018. Global- to deposit-scale controls on orthomagmatic Ni-Cu (-PGE) and PGE reef ore formation. Processes and Ore Deposits of Ultramafic-Mafic Magmas Through Space and Time. Elsevier, 1–46 pp.

Berning, J., Cooke, R., Heimstra, S.A., Hoffman, U., 1976. The Rössing Uranium Deposit, South West Africa. Econ. Geol. 71, 351–368.

Bertelli, M., Baker, T., Cleverley, J.S., Ulrich, T., 2009. Geochemical modelling of a Zn–Pb skarn: constraints from LA–ICP–MS analysis of fluid inclusions. J. Geochem. Explor. 102, 13–26.

Bodnar, R.J., Lecumberri-Sanchez, P., Moncada, D., Steele-MacInnis, M., 2014. 13.5-Fluid inclusions in hydrothermal ore deposits. In: Turekian, H.D.H.K. (Ed.), Treatise on Geochemistry, second ed Elsevier, Oxford, 119–142 pp.

Brugger, J., Liu, W., Etschmann, B., Mei, Y., Sherman, D.M., Testemale, D., 2016. A review of the coordination chemistry of hydrothermal systems, or do coordination changes make ore deposits? Chem. Geol. 447, 219–253. Available from: https://doi.org/10.1016/j.chemgeo.2016.10.021.

Casa, G.D., Manni, A., Saviano, G., Violo, G., 2003. Gold occurrence in Central Italy—the Ponte San Pietro mineralization. Ore Geol. Rev. 23, 99–105. Available from: https://doi.org/10.1016/S0169-1368(03)00017-9.

Černý, P., Ercit, T.S., Vanstone, P.J., 1996. Petrology and mineralization of the Tanco rare element pegmatite, southeastern Manitoba. Geological Association of Canada—Mineralogical Association of Canada, Joint Annual Meeting, Winnipeg 1996, Field Trip Guidebook A3, 63 pp.
Chivers, T., 1974. Ubiquitous trisulfur radical ion S^{-3}. Nature 252, 32–33.
Chou, I.M., 1978. Calibration of oxygen buffers at elevated P and T using the hydrogen fugacity sensor. Am. Min. 63, 690–703.
Cline, J.S., Hofstra, A.H., Muntean, J.L., Tosdal, R.M., Hickey, K., 2005. Carlin-type gold deposits in Nevada: critical geological characteristics and viable models. Economic Geology 100th Anniversary Volume. pp. 451–484.
Cooke, D.R., Simmons, S.F., 2000. Characteristics and genesis of epithermal gold deposits. Rev. Econ. Geol. 13, 221–244.
Corral, I., Cardellach, E., Corbella, M., Canals, À., Griera, A., Gómez-Gras, D., et al., 2017. Origin and evolution of mineralizing fluids and exploration of the Cerro Quema Au-Cu deposit (Azuero Peninsula, Panama) from a fluid inclusion and stable isotope perspective. Ore Geol. Rev. 80, 947–960.
Cox, D.P., Singer, D., 1986. Mineral Deposit Models. US Geological Survey Bulletin, Reston, VA.
Craig, J.R., Vaughan, D.J., 1981. Ore Microscopy and Ore Petrography. Wiley, New York.
Dai, J., Mao, J., Zhao, C., 2009. New U–Pb and Re–Os age data and the geodynamic setting of the Xiaojiayingzi Mo (Fe) deposit, western Liaoning province, Northeastern China New U–Pb and Re–Os age data and the geodynamic setting of the Xiaojiayingzi Mo (Fe) deposit, western Liaoning province, Northeastern China. Ore Geol. Rev. 35 (2), 235–244.
Dill, H.G., 2015. Pegmatites and aplites: their genetic and applied ore geology. Ore Geol. Rev. 69, 417–561. Available from: https://doi.org/10.1016/j.oregeorev.2015.02.022.
Einaudi, M., 2000. Mineral resources: assets and liabilities. In: Ernst, W.G. (Ed.), Earth Systems: Processes and Issues. Cambridge University Press, pp. 346–372.
Evans, A.M., 1993. Ore Geology and Industrial Minerals – An Introduction. Blackwell Scientific Publication, Oxford.
Evans, A.M., Moon, C.J., 2006. In: Moon, et al., (Eds.), Mineral Deposit Geology and Models, 2006. Blackwell, New York, 33–51 pp.
Evans, K.A., Reddy, S.M., Tomkins, A.G., Crossley, R.J., Frost, B.R., 2017. Effects of geodynamic setting on the redox state of fluids released by subducted mantle lithosphere. Lithos 278–281, 26–42.
Fiorentini, M.L., LaFlamme, C., Denyszyn, S., Mole, D., Maas, R., Locmelis, M., et al., 2018. Post-collisional alkaline magmatism as gateway for metal and sulfur enrichment of the continental lower crust. Geochimica et Cosmochimica Acta 223, 175–197.
Frost, B.R., Frost, C.D., 2014. Essentials of Igneous and Metamorphic Petrology. Cambridge University Press, New York.
Frost, B.R., Lindsley, D.H., Andersen, D.J., 1988. Fe–Ti oxide-silicate equilibrium: assemblages with fayalitic olivine. Am. Min. 73, 727–740.
Gaffney, J.S., Marley, N.A., 2018. Periodic table. General Chemistry for Engineers 623. Available from: https://doi.org/10.1016/b978-0-12-810425-5.09983-5.
Gammons, C.H., Williams-Jones, A.E., 1997. Chemical mobility of gold in the porphyry-epithermal environment. Econ. Geol. 92, 45–59.
González-Álvarez, I., Boni, M., Anand, R.R., 2016. Mineral exploration in regolith-dominated terrains: global considerations and challenges. Ore Geol. Rev. 73, 375–379.
Grand, S.P., 2002. Mantle shear-wave tomography and the fate of subducted slabs. Phil. Trans. R. Soc. Lon. (A3260), 2475–2491.
Groves, D.I., Vielreicher, R.M., Goldfarb, R.J., Condie, K.C., 2005. Controls on the heterogeneous distribution of mineral deposits through time. Geol. Soc. Lond. Spec. Publ. 248, 71–101.

Groves, D.I., Bierlein, F.P., Meinert, L.D., Hitzman, M.W., 2010. Iron-oxide copper gold (IOCG) deposits through Earth history: implications for origin, lithospheric setting, and distinction from other epigenetic iron oxide deposits. Econ. Geol. 105, 641–654.

Guilbert, J.M., Park, C.F., 1986. The Geology of Ore Deposits. W.H. Freeman and Co, New York, 985 pp.

Heinrich, C.A., Candela, P.A., 2014. Fluids and ore formation in the Earth's crust. In: Treatise on Geochemistry, pp. 1–28.

Holland, H.D., 2005. Sedimentary mineral deposits and the evolution of Earth's near-surface environments. Econ. Geol. 100, 1489–1509.

Hu, Z., Gao, S., 2008. Upper crustal abundances of trace elements: a revision and update. Chem. Geol. 253 (3–4), 205–221. Available from: https://doi.org/10.1016/j.chemgeo.2008.05.010.

Kazemi, K., Kananian, A., Xiao, Y., Sarjoughian, F., 2018. Petrogenesis of middle-Eocene granitoids and their Mafic microgranular enclaves in central Urmia-Dokhtar magmatic arc (Iran): evidence for interaction between felsic and mafic magmas. Geosci. Front. 10 (2), 705–723. Available from: https://doi.org/10.1016/j.gsf.2018.04.006.

Kesler, S.E., 1994. Mineral Resources, Economics and the Environment. McMillan, New York.

Kosman, C.W., Kopylova, M.G., Stern, R.A., Hagadorn, J.W., Hurlbut, J.F., 2016. Cretaceous mantle of the Congo craton: evidence from mineral and fluid inclusions in Kasai alluvial diamonds. Lithos 265, 42–56. Available from: https://doi.org/10.1016/j.lithos.2016.07.004.

Kotzer, T.G., Kyser, T.K., 1995. Petrogenesis of the proterozoic Athabasca Basin, Northern Saskatchewan, Canada, and its relation to diagenesis, hydrothermal uranium mineralization and paleohydrogeology. Chem. Geol. 120, 45–89.

Kress, V.C., Carmichael, I.S.E., 1991. The compressibility of silicate liquids containing Fe_2O_3 and the effect of composition, temperature, oxygen fugacity and pressure on their redox states. Contrib. Miner. Petrol. 108, 82–92.

Kretschmar, U., McBride, D., 2016. Understanding hydrothermal Systems. In: Kretschmar, Ulrich, McBride, Derek (Eds.), The Metallogeny of Lode Gold Deposits, pp. 151–198. First Edition Elsevier, 345p.

Kyser, K., 2016. Exploration for uranium. In: Hore Lacy, Ian (Ed.), Uranium for Nuclear Power, Resources Mining and transformation to fuel, pp. 53–76, First Edition Elsevier, 488p.

Laznicka, P., 2014. Giant metallic deposits—a century of progress. Ore Geol. Rev. 62, 259–314. Available from: https://doi.org/10.1016/j.oregeorev.2014.03.002.

Leach, D.L., Sangster, D.L., Kelley, K.D., Ross, R.L., Garven, G., Allen, C.R., 2005. Sediment-hosted Pb-Zn deposits: a global perspective. In: Hedenquist, J.W., Thompson, J.F.H., Goldfarb, R.J., Richards, J.P. (Eds.), Economic Geology 100th Anniversary, (1905–2005), pp. 5–24, Society of Economic Geologists, 1133p

Legros, H., Richard, A., Tarantola, A., Kouzmanov, K., Mercadier, J., Vennemann, T., et al., 2019. Multiple fluids involved in granite-related W-Sn deposits from the world-class Jiangxi province (China). Chem. Geol. 508, 92–115.

Liao, M., Tao, Y., Song, X., Li, Y., Xiong, F., 2016. Study of oxygen fugacity during magma evolution and ore genesis in the Hongge mafic–ultramafic intrusion, the Panxi region, SW China. Acta Geochim. 35 (1), 25–42.

Lindgren, W., 1922. A suggestion for the terminology of certain mineral deposits. Econ. Geol. 17 (4), 292–294.

Lindgren, W., 1933. Mineral Deposits, fourth ed McGraw-Hill, 930 pp.

Lindsley, D.H., 1991. Experimental studies of oxide minerals. Rev. Min. Geochem. 25, 69–106.

Macheyeki, A.S., 2011. Application of lithogeochemistry to exploration for Ni-Cu sulfide deposits in the Kabanga Belt, NW Tanzania. J. Afr. Earth Sci. 61, 62–81.
Maier, W.D., Groves, D.I., 2011. Temporal and spatial controls on the formation of magmatic PGE and Ni-Cu deposits. Mineralium Deposita 46, 841–858.
Mason-Apps, A.D., 1998. The petrology and geochemistry of the lower pyroxenite succession of the Great Dyke in the Mutorashanga area. Thesis submitted in partial fulfillment of the requirements for the Degree of Master of Science (Economic Geology) in the Department of Geology, Rhodes University, Grahamstown.
Meyer, C., 1981. Ore-forming processes in geologic history. In: Economic Geology, 75th Anniversary, pp. 6–41, The Economic Geology Publishing House, 960p.
Misra, K.C., 2000. Understanding Mineral Deposits. Kluwer Academic Publishers, p. 845.
Moon, C.J., 2016. In: Moon, et al., (Eds.), Exploration Geochemistry, 2006. Blackwell, New York, 156-198 pp.
Moon, C.J., Whateley, M.K.G., Evans, A.M., 2006. Introduction to Mineral Exploration, second ed Blackwell Publishing, New York.
Mungall, J.E., 2014. Geochemistry of magmatic ore deposits. In: Turikian, K.K., Holland, H.D. (Eds.), Treatise on Geochemistry, pp. 195–2181, Elsevier, 9144p.
Naldrett, A.J., 1997. Key factors in the genesis of Noril'sk, Sudbury, Jinchuan, Voisey's bay and other world-class Ni-Cu-PGE deposits: implications for exploration. Austr. J. Earth Sci. 44 (3), 283–315.
Naldrett, A.J., 2004. Magmatic Sulfide Deposits: Geology, Geochemistry and Exploration. Springer, New York, 730 pp.
Naldrett, A.J., 2010. From the mantle to the bank: the life of a Ni–Cu–(PGE) sulfide deposit. South Afr. J. Geol. 113, 1–32.
Niggli, P., 1929. In: Boydell, H.C. (Ed.), Ore Deposits of Magmatic Origin: Their Genesis and Natural Classification. Thomas Murby & Co, London, 93 pp.
O'Brien, H., 2015. Introduction to carbonatite deposits of Finland. Min. Depos. Finl. 291–303. Available from: https://doi.org/10.1016/b978-0-12-410438-9.00011-x.
Partington, G.A., 1990. Environment and structural controls on the intrusion of the giant rare metal Greenbushes pegmatite. West. Austr. Econ. Geol. 85, 437–456.
Peck, D.C., Huminicki, M.A.E., 2016. Value of mineral deposits associated with mafic and ultramafic magmatism: implications for exploration strategies. Ore Geol. Rev. 72, 269–298. Available from: https://doi.org/10.1016/j.oregeorev.2015.06.004.
Petrov, S.V., 2004. Economic deposits associated with the alkaline and ultrabasic complexes of the Kola Peninsula. In: Wall, V., Zaitsev, A.N. (Eds.), Phoscorites and Carbonatites from Mantle to Mine: the Key Example of the Kola Alkaline Province. Mineralogical Society, London, pp. 469–490.
Phillips, G.N., Evans, K.A., 2004. Role of CO_2 in the formation of gold deposits. Nature 429, 860–863.
Pokrovski, G.S., Dubessy, J., 2015. Stability and abundance of the trisulfur radical ion $S3^-$ in hydrothermal fluids. Earth Planet Sci. Lett. 411, 298–309.
Pun, A., Papike, J.J., Layne, G.D., 1997. Subsolidus REE partitioning between pyroxene and plagioclase in cumulate eucrites: an ion microprobe investigation. Geochimica et Cosmochimica Acta 61 (23), 5089–5097. Available from: https://doi.org/10.1016/s0016-7037(97)00295-0.
Richard, A., Cauzid, J., Cathelineau, M., Boiron, M.C., Mercadier, J., Cuney, M., 2013. Synchrotron XRF and XANES investigation of uranium speciation and element distribution in fluid inclusions from unconformity-related uranium deposits. Geofluids 13, 101–111.
Richards, J.P., 2011. Magmatic to hydrothermal metal fluxes in convergent and collided margins. Ore Geol. Rev. 40, 1–26.
Ridley, J., 2013. Ore Deposit Geology. Cambridge University Press, 409 pp.

Robb, L., 2005. Introduction to Ore-Forming Processes. Blackwell Sciences Ltd, Hoboken, NJ, 376 pp.
Rollinson, H.R., 1993. Using Geochemical Data: Longman Geochemistry Series, first ed. Routledge, 384 pp.
Rollinson, H., 2012. Geochemical constraints on the composition of Archaean lower continental crust: partial melting in the Lewisian granulites. Earth Planet. Sci. Lett. 351–352, 1–12. Available from: https://doi.org/10.1016/j.epsl.2012.07.018.
Rudnick, R.L., Gao, S., 2003. Composition of the continental crust. In: Rudnick RL (Ed.) Treatise on Geochemistry. The Crust (3), 1–64.
Sangster, D.F., 1980. Plate tectonics and mineral deposits: a view from two perspectives. Geosci. Canada 6 (4), 185–188.
Schneider H.J., 1964. Facies differentiation and controlling factors for depositional lead-zinc concentration in the Ladinian geosyncline of Eastern Europe. In: G.C. Amstutz (Ed.). Sedimentology and Ore Genesis, pp. 29–45.
Seedorff, E., Dilles, J., Proffett Jr., J.M., Einaudi, M.T., Zurcher, L., Stavast, W.J.A., et al., 2005. Porphyry deposits: characteristics and origin of hypogene features. In: Economic Geology 100th Anniversary (1905–2005). Society of Economic Geologists, pp. 251–298.
Sen, G., 2014. Petrology: Principle and Practice. Springer, Heidelberg-Berlin.
Seward, T.M., Barnes, H.L., 1997. Metal transport by hydrothermal ore fluids. In: Barnes, H. L. (Ed.), Geochemistry of Hydrothermal Ore Deposits. John Wiley & Sons, pp. 435–486.
Seward, T.M., Williams-Jones, A.E., Migdisov, A.A., 2014. The chemistry of metal transport and deposition by ore-forming hydrothermal fluids. Treat. Geochem. 29–57. Available from: https://doi.org/10.1016/b978-0-08-095975-7.01102-5.
Shvarov, Y., Bastrakov, E.H.Ch, 1999. A Software Package for Geochemical Equilibrium Modelling. User's Guide (Record 1999/25, Australian Geological Survey Organization).
Stanton, R.L., 1972. Ore Petrology. McGraw-Hill Book Co, London and New York, 713 pp.
Steudel, R., Steudel, Y., 2013. Polysulfide chemistry in sodium–sulfur batteries and related systems—a computational study by G3X(PM2) and PCM calculations. Chem. Eur. J. 19, 3162–3176.
Stowe, C.W., 1994. Composition and tectonic settings of chromite deposits through time. Econ. Geol. Bull. Soc. Econ. Geol. 89, 528–546.
Sun, W., Xing, M.L., Yong, Y.X., WeiMing, F., Xing, D., Ying, L.H., 2010. Ridge subduction and porphyry copper-gold mineralization: an overview. Sci. China, Earth Sci. 53 (4), 475–484.
Sun, W., Huang, R., Li, H., Hu, Y., Zhang, C., Sun, S., et al., 2015. Porphyry deposits and oxidized magmas: a review. Ore Geol. Rev. 65, 97–131.
Taylor, H.P., 1997. Oxygen and hydrogen isotope relationships in hydrothermal mineral deposits. In: Barnes, H.L. (Ed.), Geochemistry of hydrothermal ore deposits, third ed John Wiley & Sons, New York, pp. 229–302.
Trail, D., Bindeman, I.N., Watson, E.B., Schmitt, A.K., 2009. Experimental calibration of oxygen isotope fractionation between quartz and zircon. Geochimica et Cosmochimica Acta 73 (23), 7110–7126.
Trail, D., Watson, E.B., Tailby, N.D., 2012. Ce and Eu anomalies in zircon as proxies for the oxidation state of magmas. Geochim. Cosmochim. Acta 97, 70–87.
Tucker, R.F., Viljoen, R.P., Viljoen, M.J., 2016. A review of the Witwatersrand Basin: the world's greatest goldfield. Episodes 39 (2), 105.
Vigneresse, J.L., 2019. Addressing ore formation and exploration. Geosci. Front. 10 (4), 1613–1622.
Viter, P., Léo, A.H., Henrique, R.L., Carlos, K.J., Evaldo, S., Vinícius, R.M., et al., 2013. Geology, ore characterization and fluid inclusions study of gold deposits from the Campestre Formation, Southernmost Brazil. c02. Recursos Minerales. Montesinos de Almeida, Delia del Pilar Brasil–UFPel; UFRGS; UNIPAMPA.

Westphal, G., Kristen, G., Wegener, W., Ambatiello, P., Geyer, H., Epron, B., et al., 2010. Sodium Chloride. Wiley-VCH Verlag GmbH & Co, Weinheim.
White, W.M., Patchett, J., 1984. HfNdSr isotopes and incompatible element abundances in island arcs: implications for magma origins and crust-mantle evolution. Earth Planet Sci. Lett. 67 (2), 167–185. Available from: https://doi.org/10.1016/0012-821x(84)90112-2.
Williams, P.J., Barton, M.D., Johnson, D.A., Fontbote, L., De Haller, A., Mark, G., et al., 2005. Iron oxide copper gold deposits: geology, space-time distribution and possible modes of origin. Economic Geology 100th Anniversary Volume, 371–405.
Xiong, X., Zhu, L., Zhang, G., Li, N., Yuan, H., Ding, L., et al., 2018. Fluid inclusion geochemistry and magmatic oxygen fugacity of the Wenquan Triassic molybdenum deposit in the Western Qinling Orogen, China. Ore Geol. Rev. 99, 244–263.
Yang, Q.-Y., Santosh, M., 2015. Early cretaceous magma flare-up and its implications on gold mineralization in the Jiaodong Peninsula, China. Ore Geol. Rev. 65, 626–642. Available from: https://doi.org/10.1016/j.oregeorev.2014.01.004.
Zhou, T., Zeng, Q., Chu, S., Zhou, L., Yang, Y., 2018. Magmatic oxygen fugacities of porphyry Mo deposits in the East Xing'an-Mongolian Orogenic Belt (NE China) with metallogenic implications. J. Asian Earth Sci . Available from: https://doi.org/10.1016/j.jseaes.2018.04.004.
Zhu, Y., An, F., Tan, J., 2011. Geochemistry of hydrothermal gold deposits: a review. Geosci. Front. 2 (3), 367–374.

Further reading

Allen, P.A., 1997. Earth Surface Processes. Blackwell Science, Oxford, 404 pp.
Bianconi, F., Borshoff, J., 1984. Surficial Uranium occurrences in United Republic of Tanzania. Surficial Uranium Deposit. IAEA, Vienna, pp. 231–235.
Boniface, N., 2017. Crystal chemistry of pyrochlore from the Mesozoic Panda Hill carbonatite deposit, western Tanzania. J. Afr. Earth Sci. 126, 33–44.
Borg, G., Krogh, T., 1999. Isotopic age data of single zircons from the Archaean Sukumaland Greenstone Belt, Tanzania. J. Afr. Earth Sci. 29, 301–312.
Bowell, R.J., Barnes, A., Grogan, J., Dey, M., 2009. Geochemical Control on Uranium Precipitation in Calcrete Palae Channel Deposits of Namibia. SRK Publications, pp. 1–4.
Carlisle, D., 1984. Surficial Uranium Occurrences in Relation to Climate and Physical Setting. Surficial Uranium Deposit. IAEA, Vienna, pp. 25–35.
Chamberlain, C.M., Tosdal, R.M., 2007. U–Pb geochronology of the Lake Victoria Greenstone Terrane, Tanzania. Mineral Deposit Research Unit, The University of British Columbia. Unpublished Geita Gold Mine internal report, pp. 81.
Craig, J.R. and Vaughan, D.J., 1986. Paragenetic studies of growth-banded sphalerites in Mississippi Valley-type zinc deposits of the Appalachians. In: Augustithes, S.S. (Ed.), Mineral Paragenesis, pp. 133–158.
de Klerk, S.-L., 2001. Gold deposits in the SADC region, mineral resources survey programme No. 4, SADC Mining Sector Coordinating Unit, Council for Geoscience, 126 pp.
De Pierre, M., Coron, N., Dambier, G., Leblanc, J., Moalic, J.-P., 2003. Experimental detection of α-particles from the radioactive decay of natural bismuth. Nature 422 (6934), 876–878.
Evans, D.M., Boad, I., Byemelwa, L., Gilligan, J., Kabete, J., Marcet, P., 2000. Kabanga magmatic nickel sulfide deposits, Tanzania: morphology and geochemistry of associated intrusions. J. Afr. Earth Sci. 30 (3), 651–674.

Evans, D.M., Hunt, J.P.P.M., Simmonds, J.R., 2016. An overview of Nickel mineralization in Africa with emphasis on the Mesoproterozoic East African Nickel Belt (EANB). In: Wilson and Viljoen (2016) on Special Issue for the 35 IGC, Cape Town, South Africa.

Feneyrol, J., Giuliani, G., Ohnenstetter, D., Fallick, A.E., Martelat, J.M., Monie, P., et al., 2017. Age and origin of the tsavorite and tanzanite mineralizing fluids in the neproterozoic Mozambique Metamorphic belt. Can. Min. 55, 763–786.

Geosurvey International GMbH, 1981. Synoptic map showing the gold potential of the Archean Greenstone Belts, scale 1:500,000.

Geosurvey International GMbH, 1982. An economic appraisal of the mineral potential of the United Republic of Tanzania, Report printed for the Ministry of Energy and Minerals, Tanzania.

Giuliani, G., Ohnenstetter, D., Palhol, F., Feneyrol, J., Boutroy, E., De Boissezon, H., et al., 2008. Karelianite and vanadian phlogopite from the Merelani Hills gem zoisite deposits, Tanzania. Can. Min. 46, 1183–1194.

Haggerty, S.E., 1999. A diamond trilogy: super plumes, supercontinents and supernovae. Science 285, 851–860.

Haldar, S.K., 2013. Mineral Exploration: Principles and Applications. Elsevier, China.

Haldar, S.K., 2017. Platinum-Nickel-Chromium Deposits: Geology, Exploration and Reserve Base. Elsevier, 322p.

Harland, W.B., Armstrong, R.L., Cox, A.V., Craig, L.E., Smith, A.G., Smith, D.G., 1990. A Geological Time Scale, 1989. Cambridge University Press, 263 pp.

Harris, F.J., 1961a. Summary of the Geology of Tanganyika. Part IV, Economic Geology Memoir No. 1. Dodoma Government Printer.

Harris, J.F., 1961b. Summary of the Geology of Tanzania, Economic Geology. Tanzania Geological Survey, Dodoma, Tanzania.

Harris, F.J., 1969. Kaolin deposit of Tanzania, international Geological Congress Report of the twenty third session, Czechslovakia, 1969, Proceeding of the Symposium 1. Kaolin deposit of the world, B-Oversea Contributions, 75–78.

Heirtzler, J.R., Dickson, G.O., Herron, E.M., Pitman, W.C., Le Pichon, X., 1968. Marine magnetic anomalies, geomagnetic field reversals and motions of the ocean floor and continents. J. Geophys. Res. 73, 2119–2136.

ICS 2000 edition of the International Stratigraphic Chart published and sanctioned by the ICS and IUGS. <http://www.angelfire.com/sc3/farooqs/notes/b3-3-1.htm>.

Jansson, N.F., Zetterqvist, A., Allen, R.L., Billström, K., Malmström, L., 2017. Genesis of the Zinkgruvan stratiform Zn-Pb-Ag deposit and associated dolomite-hosted Cu ore, Bergslagen, Sweden. Ore Geol. Rev. 82, 285–308.

Kabete, J.M., Groves, D.I., McNaughton, N.J., Mruma, A.H., 2012. A new tectonic and temporal framework for the Tanzanian Shield: implications for gold metallogeny and undiscovered endowment. Ore Geol. Rev. 48, 88–124.

Kazimoto, E.O., Schenk, V. and Appel, P., 2015. The age of Au–Cu–Pb-bearing veins in the poly-orogenic Ubendian Belt (Tanzania): U–Th–total Pb dating of hydrothermally altered monazite, In: Contributions to Mineralogy and Petrology *Petrology*. Springer-Verlag Berlin Heidelberg 169, 1088.

Keller, W.D., 1978. Classification of kaolins exemplified by their textures in scan electron micrographs. Clays Min. 26, 1–20.

Lawley, C.J.M., Selby, D., Condon, D.J., Horstwood, M., Millar, I., Crowley, Q., et al., 2014. Lithogeochemistry, geochronology and geochronology and geodynamic setting of the Lupa Terrane, Tanzania: implications for the Archean Tanzania Craton. Precambrian Res. 231, 174–193.

Li, Y., Audétat, A., 2013. Gold solubility and partitioning between sulfide liquid, monosulfide solid solution and hydrous mantle melts: implications for the formation of Au-rich

magmas and crust–mantle differentiation. Geochimica et Cosmochimica Acta 118, 247–262.
Li, Z.X., Bogdanova, S.V., Collins, A.S., Davidson, A., Waele, B. De, Ernst, R.E., et al., 2008. Assembly, configuration, and break-up history of Rodinia: a synthesis. Precambrian Res. 160, 179–210.
Maier, W.D., Barnes, S.J., de Waal, S.A., 1998. Exploration for magmatic Ni-Cu-PGE sulphide deposits: a review of recent advances in the use of geochemical tools, and their application to some South African Ores. South Afr. J. Geol. 101 (3), 237–253.
Malisa, E., Mhuhongo, S., 1990. Tectonic setting of gemstone mineralization in the Proterozoic metamorphic terrane of the Mozambique Belt in Tanzania. Precambrian Res. 46, 167–176.
McKinlay, A.C.M., 1965. The Coalfields and Coal Resources of Tanzania. Tanzania Geological Survey, Dodoma, Tanzania, Bulletin No. 38.
McQueen, K.G., 1987. Deformation and remobilization in some Western Australian nickel ores. Ore Geol. Rev. 2 (1–3), 269–286.
McQueen, K.G., 2006. Geochemical fractionation in the regolith of western New South Wales. Geochimica et Cosmochimica Acta 70 (18), A415.
MEM, 2005: TANZANIA: Opportunities for Mineral Resources Development, Ministry of Energy and Minerals, Dares-Salaam, Tanzania.
Mtelela, C., 2016. Sedimentology and Stratigraphy of the Late Cenozoic Lake Beds Succession, Rukwa Rift Basin, Tanzania: Implications for Hydrocarbon Prospectivity (PhD thesis). James Cook University.
Nance, R.D., Worsley, T.R., Moody, J.B., 1986. Post-Archean biogeochemical cycles and long-term episodicity in tectonic processes. Geology 14, 514–518.
Parnell, J., Ye, L., Chen, C., 1990. Sediment-hosted mineral deposits. Special Publication, 11, International Association of Sedimentologists. Blackwell Scientific Publications, Oxford, 227 pp.
Pearson, R.G., 1963. Hard and soft acids and bases. J. Am. Chem. Soc. 85, 3533–3539.
Reeves, C.V., de Wit, M.J., Sahu, B.K., 2004. Tight reassembly of Gondwana exposes Phanerozoic shears in Africa as global tectonic players. Gondwana Res. 7 (I), 7–19.
Richard, A., 2017. Radiolytic (H_2, O_2) and other trace gases (CO_2, CH4, C_2H_6, N_2) in fluid inclusions from unconformity-related U deposits. 15th Water-Rock Interaction International Symposium, WRI-15. Procedia Earth and Planetary Science 17, 273–276.
Richard, A., Banks, D.A., Hendriksson, N., Lahaye, Y., 2018. Lithium isotopes in fluid inclusions as tracers of crustal fluids: an exploratory study. J. Geochem. Explor. 184, 158–166.
Richards, J.P., 2013. Giant ore deposits formed by optimal alignments and combinations of geological processes. Nat. Geosci. 6, 911–916.
Roedder, E., 1976. Fluid-inclusion evidence on the genesis of ores in sedimentary and volcanic rocks. In: Wolf, K.H. (Ed.), Handbook of Stratabound and Stratiform Ore Deposits., 2. Elsevier, Amsterdam, pp. 67–110.
Roedder, E., 1979. Fluid inclusions as samples of ore Fluids. In: Barnes, H.L. (Ed.), Geochemistry of Hydrothermal Ore Deposits, second ed. Wiley-Interscience, New York, 614-737 pp.
Sango, P.M., 1988. Structural and Lithological controls of gold mineralization in the Lupa goldfields, Tanzania. In: Ho, S.E., Groves, D.I. (Eds.), Recent Advances in understanding Precambrian Gold Deposits, 12. Geology Department and University Extension, The University of Western Australia, Publication, pp. 99–109.
Sanislav, I.V., Wormald, R.J., Dirks, P.H.G.M., Blenkinsop, T.G., Salamba, L., Joseph, D., 2014a. Zircon U-Pb ages and Lu-Hf isotope systematic from late-tectonic granites, Geita greenstone belt: implications for crustal growth of the Tanzania craton. Precambrian Res. 242, 187–204.

Sanislav, I.V., Dirks, P.H.G.M., Cook, Y.A., Blenkinsop, T.G., Kolling, S.L., 2014b. A Giant Gold System, Geita Greenstone Belt, Tanzania. ACTA Geologica Sinica (English Edition) 88 (Suppl), 2.

Sanislav, I.V., Kolling, S.L., Brayshaw, M., Cook, Y.A., Dirks, P.H.G.M., Blenkinsop, T.G., et al., 2015. The geology of the giant Nyankanga gold deposit, Geita Greenstone Belt, Tanzania. Ore Geol. Rev. 69, 1–16.

Sanislav, I.V., Brayshaw, M., Kolling, S.L., Dirks, P.H.G.M., Cook, Y.A., Blenkinsop, T.G., 2016. The structural history and mineralization controls of the world-class Geita Hill gold deposit, Geita Greenstone Belt, Tanzania. Miner Deposita 52 (2), 292–315. Available from: https://doi.org/10.1007/s00126-016-0660.

Schwaighofer, B., Muller, H.W., 1987. Mineralogy and genesis of the Pugu Hill Kaolin. Clay Min. 22, 401–409.

Selley, R.C., 1998. Elements of Petroleum Geology, second ed Academic Press, New York, 470 pp.

Smith, W., Sandwell, D., 1997. Measured and estimated seafloor topography (version 4.2). World Data Center A for Marine Geology and Geophysics Res. publ. RP-1, poster, 34" x 53".

Song, X., Wang, Y., Chen, L., 2011. Magmatic Ni-Cu-(PGE) deposits in magma plumbing systems: features, formation and exploration. Geosci. Front. 2 (3), 375–384.

Stein, M., Hofmann, A.W., 1994. Mantle plumes and episodic crustal growth. Nature 372, 63–67.

Stiefenhofer, Farrow, 2004. Geology of the Mwadui kimberlite, Shinyanga district, Tanzania. Lithos 76 (1), 139–160.

Sun, W., 2017. The initiation of plate subduction. Solid Earth Sci. 2 (4), 89–90. Available from: https://doi.org/10.1016/j.sesci.2017.10.001.

Switzer, G.S., 1974. Composition of green garnet from Tanzania and Kenya. Gems Gemol. 14, 296–297.

Tack, L., Wingate, M.T.D., De Waele, B., Meert, J., Belousova, E., Griffin, B., et al., 2010. The 1375 Ma "Kibaran event" in Central Africa: prominent emplacement of bimodal magmatism under extensional regime. Precambrian Res. 180, 63–84.

Thomas, R.J., Spencer, C., Bushi, A.M., Baglow, N., Boniface, N., Kock, G., et al., 2016. Geochronology of the central Tanzania Craton and its southern and eastern orogenic margins. Precambrian Res. 277, 47–67.

Torsvik, T.H., Smethurst, M.A., Meert, J.G., Van der Voo, R., McKerrow, W.S., Sturt, B.A., et al., 1996. Continental break-up and collision in the Neoproterozoic and Palaeozoic – a tale of Baltica and Laurentia. Earth-Sci. Rev. 40, 229–258.

White, W.M., 2013. Geochemistry, first ed Wiley-Blackwell, Oxford, 660 pp.

White, W.M., Klein, E.M., 2014. Composition of the oceanic Crust. Treat Geochem. 457–496. Available from: https://doi.org/10.1016/b978-0-08-095975-7.00315-6.

Windley, B.F., 1995. The Evolving Continents, third ed. John Wiley & Sons, New York, Brisbane, Toronto, Singapore, xvii + 526 pp. Chichester.

Xu, C., Wang, L., Song, W., Wu, M., 2010. Carbonatites in China: a review for genesis and mineralization. Geosci. Front. 1 (1), 105–114. Available from: https://doi.org/10.1016/j.gsf.2010.09.001.

CHAPTER

3

Conventional and nonconventional exploration techniques—principles

The driving forces for invention of nonconventional exploration techniques are many. However, one of which is the global population versus mineral resources. Mineral deposits are decreasing exponentially against geometrically increasing global population needs. Ordinary means of searching for minerals (conventional ways) are becoming less and less popular because their power to discover concealed ore deposits is limited (e.g., Kesler, 2000; Xiong et al., 2018). The alternative solution is to apply nonconventional exploration techniques. The concept of conventional techniques and the need to have nonconventional techniques do not only apply to the mineral industry but to all-natural resources including oil and gas (e.g., Moon and King, 2015; Seneshen, 2015). In this chapter, selected conventional and nonconventional techniques have been described.

3.1 Conventional techniques

Conventional techniques are traditional techniques. In this context, they are ordinary methods for searching minerals. For instance, if someone is searching for gold using soil or stream sediment surveys by analyzing soil or sediments for gold only—it is a conventional way. In other words, if there is gold in the soil or stream sediments, the anomaly can be picked and be followed up by detailed soil/stream sediment sampling or drilling. However, it is possible that the anomaly trails could be too feeble to have moved a wider area such that soil or stream sediments could not pick the geochemical halos of the gold deposit by simply analyzing gold in the soil or in the stream sediments. The feebleness

DOI: https://doi.org/10.1016/B978-0-12-819495-9.00003-7

© 2020 Elsevier Inc. All rights reserved.

of geochemical halos could have been attributed to (1) presence of geochemical barriers such as duricrusts that hinder geochemical halos from the ore underneath to reach the soil on top and (2) relatively deep burial of ore deposit by rocks or by overburden (e.g., Cameron et al., 2002; Anand et al., 2016), among other factors.

3.1.1 Geochemical anomaly

Mineral explorers have used various ways in the past to locate mineral deposits. One of the key techniques has been vested in the establishment of geochemical anomalies. A geochemical anomaly can be generally defined as a zone or an area of abnormal high or low values of element(s) or mineral(s) in question as compared to the overall element(s) background values. Conventional methods usually focus on the detection of positive geochemical anomalies and ignore negative geochemical anomalies related to mineralization (Xiong et al., 2018; Zuo and Xiong, 2018). The abnormal concentration of elements or minerals is usually picked from the soils, waters, sediments, or rocks lying either vertically above deposits or far away from the deposit(s). Table 3.1 shows background concentration of trace elements from upper crustal samples that include (1) the suite of Post-Archean Shales (PAAS), which Taylor and his coworkers originally used to constrain the rare earth element contents of the upper crust; (2) SCO-1, a shale standard reference material from the USGS; (3) worldwide loess, and (4) upper crustal composites from China that include graywackes, shales, and granites (Hu and Gao, 2008) (Fig. 3.1).

Literally, any highest or lowest values in a dataset would imply the presence of anomaly(s), that is, something abnormal. The highest values would represent mineralization zones and the lowest values would indicate mineral alterations (e.g., McIlwaine et al., 2016, 2017; Parsapoor et al., 2017). Unfortunately, such an approach may lead to frustrations because not all highest values represent mineralized zones. They could just be representing upper limits of background values for any given area because background values are not a single value but a range of values. Further, abnormally higher values are not necessarily caused by economic mineralization but could be due to geogenic and anthropogenic processes (Albanese et al., 2007). Technically, an anomaly is a zone defined by values that are above threshold values.

A threshold value is a value above which anomalous values emerge—it is the average of upper limit of the background values. Ideally, rocks of the same chemical composition would have same background values and therefore same threshold values for specific elements or minerals reflecting underlying parent rocks (e.g., Fernandes et al., 2018).

TABLE 3.1 Analyses of forty-eight trace elements in five international rock standards (ppm) and blanks (ppb).

Ppm	AGV-1 (andesite)			BHVO-1 (basalt)			G-2 (granite)			SCO-1 (shale)			GSR-5 (shale)			Blank	
	Ref	$N = 6$	RSD%	Ref	$N = 5$	RSD%	Ref	$N = 7$	RSD%	Ref	$N = 4$	RSD%	Ref	$N = 4$	RSD%	$N = 5$	STD
Li	10.7	11.0	4.44	4.6	5.13	2.78	34	33.1	6.15	45	48.4	4.79	44	55.0	3.77	0.036	0.028
Be	2.1	2.22	3.78	1.1	0.97	4.53	2.5	2.60	5.77	1.84	1.91	3.44	3.0	3.16	3.25	0.00091	0.00049
B	8.1	8.07	6.96	2.5	3.58	15.2	2.4	2.49	24.3	72	76.0	6.99	154	159	5.21	0.39	0.26
Sc	12.3	12.2	4.68	31	31.0	4.79	3.5	3.89	5.47	10.8	11.8	3.32	18.5	19.2	5.54	0.033	0.011
V	119	123	7.17	318	318	3.20	36	36.6	3.96	131	132	3.05	87	91.4	4.03	0.50	0.38
Cr	9.4	9.81	6.41	287	296	3.70	8.7	8.20	5.88	68	72.5	4.26	99	109	2.51	0.53	0.16
Co	15.2	15.8	4.76	45	44.1	2.72	4.6	4.50	2.52	10.5	11.2	4.42	21	22.0	4.64	0.0046	0.0032
Ni	15.5	16.4	2.85	118	124	4.07	(5	2.31	7.67	27	27.5	5.98	36.8	40.8	4.16	0.065	0.037
Cu	58	60.5	5.71	137	137	4.08	11	11.1	4.65	28.7	28.9	5.11	42	44.2	2.49	0.072	0.022
Zn	87	89.3	2.62	106	106	4.58	86	82.9	2.11	103	97.1	1.54	55	54.0	2.16	0.80	0.56
Ga	20.2	20.6	4.22	21	21.0	4.31	23	23.1	3.16	15	16.3	2.64	25.6	25.4	2.88	0.0037	0.0020
Ge	1.25	1.24	2.78	1.6	1.67	1.81	1.14	1.16	3.01	(1	1.72	1.57	3.1	3.55	1.35	0.0081	0.0040
As	0.88	1.08	4.19	0.5	0.66	4.09	0.25	0.28	9.74	12.4	12.2	4.00	1.4	1.54	2.19	0.021	0.009
Rb	66.6	69.9	3.07	9.19	10.0	3.16	170	168	4.49	112	110	1.95	205	207	4.08	0.11	0.05
Sr	660	671	1.60	396	402	3.90	478	475	1.12	174	165	1.26	90	91.3	2.01	0.018	0.018
Y	19	20.5	4.37	26	27.4	3.91	11	10.1	1.67	26	24.3	1.12	26	28.3	2.37	0.0018	0.0017
Zr	231	227	2.55	174	166	3.96	309	319	5.27	160	159	3.65	96	95.7	1.07	0.0030	0.0015
Nb	14.6	14.7	1.19	18.6	19.0	2.69	12	12.4	1.86	11	12.0	1.06	14.3	13.5	0.44	0.0012	0.0012

(Continued)

TABLE 3.1 (Continued)

Ppm	AGV-1 (andesite)			BHVO-1 (basalt)			G-2 (granite)			SCO-1 (shale)			GSR-5 (shale)			Blank	
	Ref	$N = 6$	RSD%	Ref	$N = 5$	RSD%	Ref	$N = 7$	RSD%	Ref	$N = 4$	RSD%	Ref	$N = 4$	RSD%	$N = 5$	STD
Mo	2.1	2.17	3.35	1	1.18	5.10	(1.1	0.28	6.67	1.37	1.27	6.06	0.35	0.42	5.64	0.035	0.030
Cd	0.061	0.062	6.11	0.094	0.10	10.9	0.016	0.014	17.3	0.14	0.16	6.19	0.033	0.0034	42.3	0.0026	0.0015
In	0.042	0.047	9.37	(0.18	0.095	10.6	0.032	0.028	5.50	(.11	0.062	4.64	0.082	0.077	2.96	0.00016	0.00004
Sn	4.9	4.83	4.33	2.1	2.01	4.02	(1.8	1.73	5.69	3.7	3.34	2.84	2.0	2.25	2.13	0.0083	0.0016
Sb	4.2	4.45	3.17	0.15	0.17	6.02	0.07	0.063	6.88	2.5	2.29	3.92	0.17	0.19	6.30	0.010	0.006
Te	0.0016	0.0034	20.3	0.0064	0.0080	6.44	(.005	0.0081	6.80	(.077	0.076	3.13	0.022	0.014	7.97	0.00097	0.00014
Cs	1.26	1.31	3.33	0.101	0.11	6.95	1.34	1.37	3.19	7.8	7.91	2.53	14	14.8	2.73	0.00060	0.00034
Ba	1200	1230	2.64	133	140	2.99	1882	1885	2.95	570	580	1.91	450	441	3.11	0.075	0.063
La	38.2	38.7	1.51	15.5	16.0	1.38	89	86.4	3.65	29.5	29.5	2.47	62	58.9	3.01	0.0025	0.0020
Ce	67.6	68.8	2.51	38.1	39.4	4.68	160	158	3.20	62	55.9	3.15	109	107	3.49	0.0028	0.0016
Pr	8.3	8.58	3.65	5.42	5.72	1.72	18	17.1	4.37	6.6	7.04	5.24	13.6	13.8	4.48	0.00056	0.00042
Nd	31.7	33.3	2.96	24.7	26.1	2.82	55	54.8	4.01	26	26.8	3.64	48	49.9	3.15	0.0019	0.0017
Sm	5.72	5.96	2.20	6.12	6.36	2.23	7.2	7.27	2.44	5.3	5.15	2.68	8.4	8.51	2.30	0.00079	0.00036
Eu	1.58	1.69	3.50	2.09	2.09	4.11	1.4	1.47	3.39	1.19	1.10	3.57	1.7	1.68	2.30	0.00023	0.00014
Gd	4.7	4.94	4.63	6.33	6.28	2.18	4.3	4.68	2.97	4.6	4.37	4.27	6.7	6.88	3.26	0.00063	0.00054
Tb	0.69	0.70	3.26	0.96	1.00	4.08	0.48	0.52	3.68	0.7	0.71	1.75	1.02	1.00	3.64	0.00013	0.00005
Dy	3.55	3.69	2.43	5.31	5.51	1.84	2.4	2.39	3.83	4.2	4.12	3.62	5.1	5.37	0.54	0.00058	0.00038
Ho	0.68	0.70	4.97	0.98	1.05	3.17	0.4	0.37	2.68	0.97	0.84	3.16	0.98	1.04	1.25	0.00013	0.00008

Er	1.82	1.76	3.14	2.55	2.53	4.41	0.92	0.96	3.60	2.5	2.26	3.78	2.7	2.74	2.27	0.00027	0.00020
Tm	0.28	0.27	5.39	0.33	0.35	4.83	(.18	0.13	4.58	0.42	0.37	4.83	0.43	0.42	2.05	0.00011	0.00003
Yb	1.63	1.72	6.47	2	2.15	3.46	0.8	0.76	3.65	2.27	2.39	2.81	2.6	2.71	2.79	0.00043	0.00025
Lu	0.244	0.27	5.27	0.27	0.31	3.42	0.11	0.11	3.11	0.34	0.37	2.87	0.41	0.42	2.98	0.00015	0.00007
Hf	5.1	5.06	3.41	4.46	4.53	3.33	7.9	8.05	2.89	4.6	4.65	3.78	2.9	2.92	4.16	0.0010	0.0012
Ta	0.87	0.92	2.19	1.21	1.29	2.48	0.88	0.87	3.90	0.92	0.90	2.70	1.0	0.94	3.16	0.00017	0.00009
W	0.52	0.57	7.96	0.21	0.24	4.72	(.2	0.11	16.0	1.4	1.55	4.93	0.79	1.10	6.20	0.023	0.010
Tl	0.35	0.33	5.74	0.044	0.045	6.19	0.91	0.92	2.95	0.72	0.74	3.09	0.71	0.68	4.65	0.0026	0.0009
Pb	37.4	39.6	5.10	2.4	2.13	5.17	30	31.6	4.15	31	31.5	4.71	8.7	8.16	3.14	0.043	0.020
Bi	0.05	0.051	5.98	0.016	0.012	10.4	0.037	0.037	3.10	0.37	0.40	2.96	0.23	0.23	2.09	0.00049	0.00028
Th	6.4	6.38	4.24	1.23	1.25	3.27	24.7	25.0	2.53	9.7	9.44	1.79	12.8	12.9	2.60	0.00062	0.00050
U	1.93	1.98	2.56	0.409	0.44	5.10	2.07	2.02	4.87	3	3.02	2.01	1.5	1.60	2.48	0.00027	0.00025

From Hu, Z., Gao, S., 2008. Upper crustal abundances of trace elements: a revision and update. Chem. Geol. 253(3–4), 205–221. doi:10.1016/j.chemgeo.2008.05.010
The RSD is the relative standard deviation in percent. The reference values (Ref) of AGV-1 and BHVO-1 are taken from the preferred values in GeoRem (http://georem.mpch-mainz.gwdg.de/). The reference values of G-2, GSR-5, and SCO-1 are taken from Govindaraju (1994). Data underlined are recommended values. Other values are proposed except those preceded by a "(" which are information values.

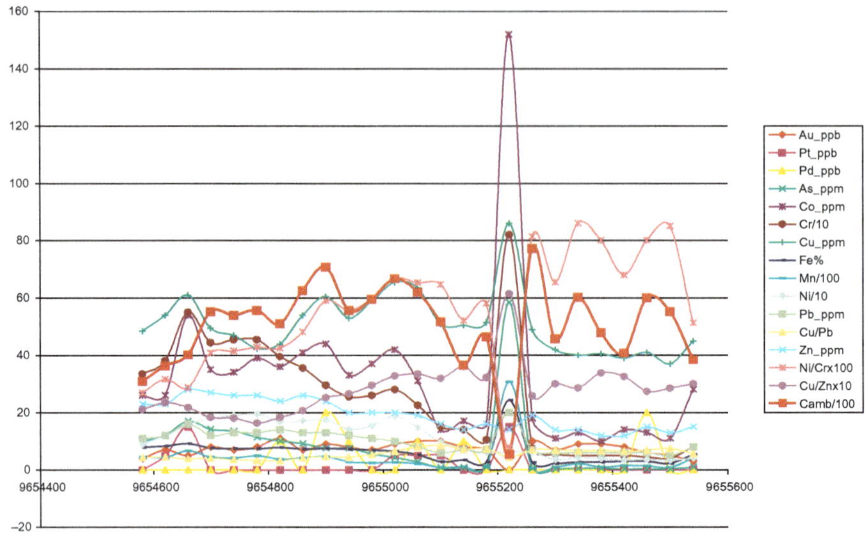

FIGURE 3.1 Geochemical anomaly (over a zone of about 50 m wide) of various elements; Pt, As, Co, Cu, etc. Note also the anomaly of element ratios and negative anomaly of Au.

Similarly, the soil lying on top of those rocks would have background values reflecting the mineralogical composition of the underlying rocks. In practice, this is far from being real. Rocks of the same nature (say granites) are by no means going to have same background values of Ni, Cu, Zn, Pb, etc. in all places. Similarly, basalts, dacites, dunites, etc. will have different background values for different elements at different localities. Attempts have been made by various authors such as Clarke (1924), Goldschmidt (1937), Fletcher (1981), and Rollinson (1993) to establish background values of various elements in different rocks and soils (see also Table 3.1).

The most important thing for explorers therefore is to undertake at early stages of mineral exploration establishment of average background values of elements of interest for a given locality—a process known as "orientation survey." The values that will be obtained from such a survey are then used for establishing threshold values of the given area.

Conventionally, threshold values have been estimated using frequency-based models such as

$$\text{Threshold} = \text{Mean} + 2 \times \text{ standard deviation} \tag{3.1}$$

(Aitchison, 1986; Filzmoser and Hron, 2008; Reimann et al., 2002). According to Sinclair (1974) however, such an attempt would complicate the results because it puts together anomalous values with background values without taking into account compositional nature of geochemical data (Zuo et al., 2015), and does ignore the fact that anomalous and background populations have fairly extensive ranges of overlap in some cases (Zuo et al., 2015). The use of Eq. (4.1) in determining threshold values, and hence anomalies, has led to some ore deposits go unnoticed in many parts of the world (Personal Communication). Of recent, percentiles have been used to refine the establishment of threshold values and the equation for threshold value is:

$$\text{Threshold} = Q3 + (Q3 - Q1) \times 1.5 \tag{3.2}$$

where $Q3$ and $Q1$ are the 75th and 25th percentiles of the element concentrations respectively (Sinclair, 1974; Jarva et al., 2010). In conventional techniques, threshold values can also be computed as median of a variable or element plus two times the median absolute deviation (MMAD) (Reimann et al., 2002; Reimann et al., 2015; Shuguang et al., 2015). Further, in the application of conventional statistical methods, there is no incorporation of the spatial pattern characteristics of shape, extent, and magnitude of anomalous areas (Cheng et al., 1996; Davis, 2002; Afzal et al., 2010; Bai et al., 2010). To overcome this, and to optimize chances to locate mineral deposits, nonconventional techniques are gaining popularity.

3.2 Nonconventional techniques

As our earth has experienced decreasing mineral resources during the human life span especially since the beginning of the industrialization era, the demand for the minerals is equally increasing geometrically. There is gambling for the limited mineral resources in almost all continents (e.g., González-Álvarez et al., 2016). This is attributed to the fact that the world population is also increasing geometrically. Presence of minerals in some parts of the world has caused disasters, instead of bringing wealth or economic prosperity and social welfare following societal struggles to get some shares from the decreasing mineral commodities. For the same reasons, therefore minerals are no longer available in quantities as they used to be in the past. Consider the fact that in the last few decades, gold nuggets would be easily found on the earth's surface, particularly in many parts of Africa and Latin America. Diamonds, tanzanite, and other currently expensive gemstones were also treated as worthless materials by the local people. Similar case applied to hydrocarbons where oil seepages/spills were commonly seen on the shores of oceans—this is seldom the case today. The demand for the minerals in the societies today is so high and the prices of the same are consequently high. However, as per Fig. 3.2, the reserves for most

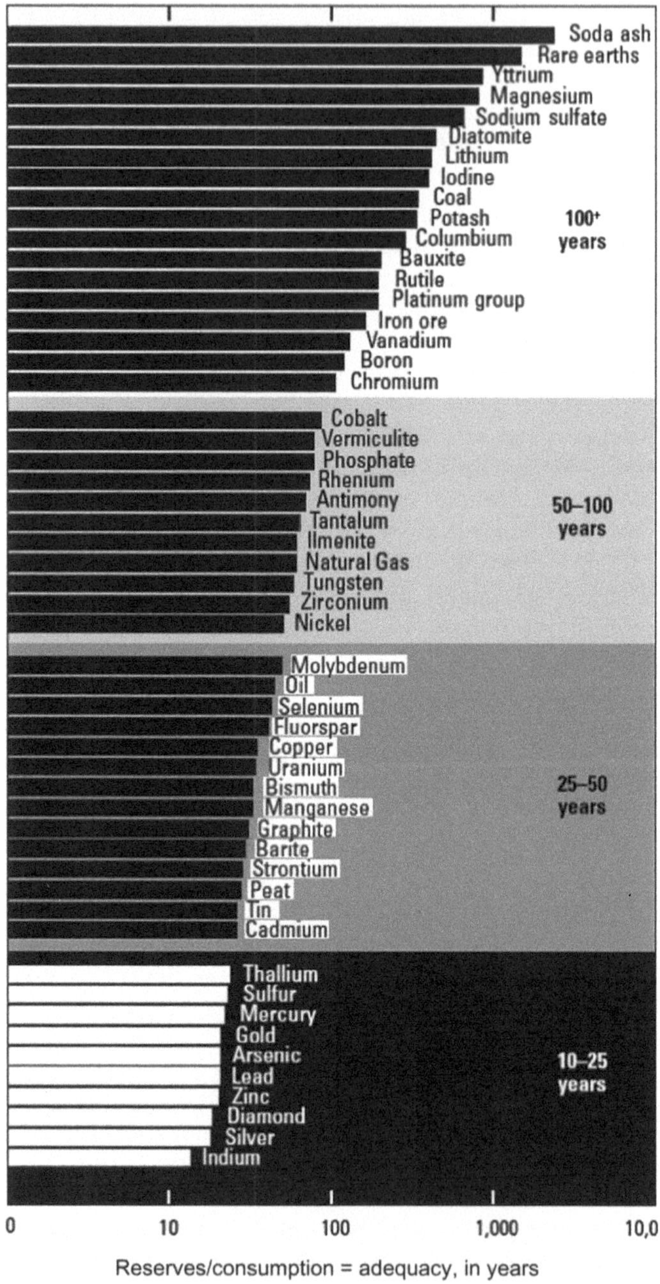

FIGURE 3.2 Global reserves to annual global consumption for mineral and energy commodities in 1992. Source: *From Wellington, T.-A.A., Mason, T.E., 2014. The effects of population growth and advancements in technology on global mineral supply. Resour. Policy 42, 73–82. doi:10.1016/j.resourpol.2014.10.006.*

of the minerals are still much higher (20–100 times) than the global consumption, thus only improved science and technology to explore the minerals is need (e.g., Kesler, 2000, 2007).

Furthermore, most or all of the near-surface ore bodies have been mined and locating deep-seated ore bodies is no longer a direct endeavor, particularly on deeply weathered terrains where ore bodies are much deeper from the surface (e.g., González-Álvarez et al., 2016). It requires sophisticated techniques of applied geophysics and applied geochemistry, among other techniques to explore these decreasing resources. This chapter highlights some of those sophisticated techniques to elucidate the power of nonconventional exploration techniques: predictive mapping, lithogeochemistry, factor analysis and principal component analysis (PCA), mineral zonations, remote sensing geochemistry, fractal models, and concentration gradient models. All these techniques are aimed at obtaining reliable geochemical anomalies (ore deposits). The techniques are effective methods for delineating mineralized zones from barren host rocks in geochemical exploration (e.g., Xiang et al., 2018). For applied geophysics, the reader is referred to Telford et al. (1990) and Parasnis (1996) for details.

3.2.1 Mineral prospectivity mapping

Mineral prospectivity mapping (also known as mineral potential mapping) is a method for determining locations where a mineral deposit is most likely to occur in a study area. The method employs geographic information system (GIS) to integrate multiple geoscience datasets followed by knowledge-driven, and data-driven mathematical tools such as weights-of-evidence, artificial neural networks, logistic regression, and fuzzy logics (Porwal and Kreuzer, 2010; Maepa and Smith, 2017; Li et al., 2019). Mineral prospectivity mapping is an important aspect of mineral exploration that is gaining popularity (Carranza and Laborte, 2015)—a method that employs a combination of various methods; geology, geochemistry, geophysics, etc. to predict the potentiality of a given area being likely to have an ore body (and at what confidence level) or not (e.g., Bagas et al., 2017). Mineral prospectivity mapping should not be confused with remote predictive mapping which is essentially the geological mapping using remotely sensed data (e.g., Schetselaar et al., 2008). Mineral predictivity mapping should also not be confused with soil predictivity papping (Florinsky, 2016) which deals with soil classification.

Mineral prospectivity mapping has been applied in various parts of the world including French Guiana by Bureau de Recherches

Géologiques et Minières (BRGM) (e.g., Cassard et al., 2008), China (Bagas et al., 2017), United States (e.g., Regmi and Rasmussen, 2018) and to a lesser extend to some other parts of Africa such as Rwanda (Personal Communication). It has also been introduced at the Geological Survey of Tanzania and some geologists in many parts of the world are aware of the power of mineral prospectivity mapping. The required data/information will largely depend on the type of the mineral or group of minerals being searched for and the associated deposit type(s). For example, if one is looking for Ni–Cu sulfides, then mafic/ultramafic rocks/and thick laterites form integral part of the important information. For gold in Green Stone Belts (GSB) for instance, such as those in Geita Tanzania (Geita GSB), presence or absence of banded iron formations (BIF), quartz veins, shear zones, faults, age (e.g., 2.5–2.7 Ga) of the rocks and structures, are some of the important information in the Geita GSB. Gold in the Geita GSB is said to be hosted in BIF, quartz veins, shear-zones and in contacts between BIF and tuff (e.g., Borg and Krogh, 1999).

Such important information (or as often called criteria) is stored as layers in a GIS environment. A good example is the work of Cassard et al. (2008) in search for gold in Precambrian orogenic setting. They use geological criteria (Lithostratigraphy, stratigraphy contacts), geophysical criteria and geochemical criteria in order to establish favorability map for French Guiana. Basically, the polygons of geological formations and lines of lithologic contacts, faults, shear zones, etc. are the key factors to establish such kinds of maps. Similar approaches have been used successfully elsewhere in the world (e.g., Cassard et al., 2001; Billa et al., 2002; Roy et al., 2006; Carranza and Laborte, 2015).

With the application of mineral prospectivity mapping software, one can be able to visualize a map showing mineral favorable areas, i.e., a map produced after all criteria have been processed. A typical map would be as that of n Fig. 3.3 where the area with highest scores is shown by red color represents the highest likelihood for finding the mineral of interest, in this case, gold.

Between 2007 and 2013, and following the evolving power of mineral prospectivity mapping, China, through the China National Mineral Assessment Initiative, undertook a national-wide initiative to compile mineral prospectivity maps and estimated the total (known and unknown) resources available in the country (Xiao et al., 2017). The authors report that through mineral prospectivity mapping, they were able to establish mineral deposit models currently applied to 1000 known deposits in China related to supergene, sedimentary and hydrothermal deposits. An example of the workflow and matrix involved in mineral prospectivity mapping (using the geological information

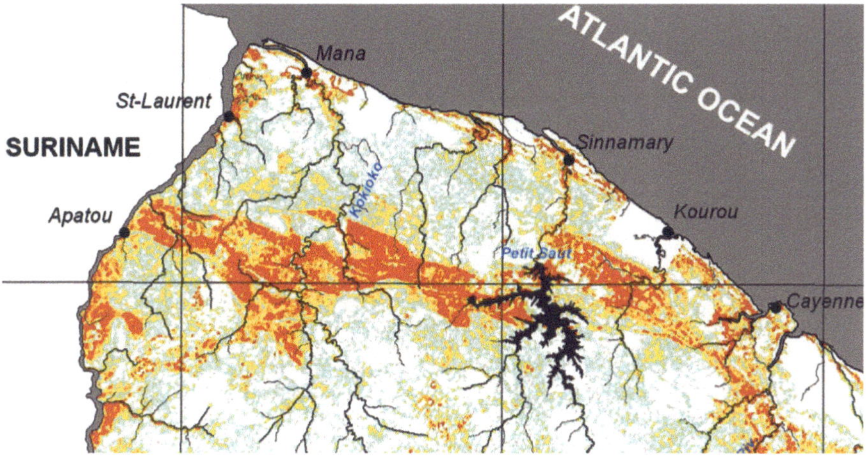

FIGURE 3.3 Favorability map (mineral prospectivity mapping) for an area in French Guiana. Note that reddish color represents areas with highest favorability (after all criteria have been processed) for a mineral in question. Source: *After Cassard, D., Billa, M., Lambert, A., Picot, J.-C., Husson, Y., Lasserre, J.-L., et al., 2008. Gold predictivity mapping in French Guiana using an expert-guided data-driven approach based on a regional-scale GIS. Ore Geol. Rev. 34, 471-500.*

mineral prediction method) in China in the framework of Project (No. 1212011120140) of the aforesaid China National Mineral Assessment Initiative is outlined by (Bagas et al., 2017, their Fig. 1; Xiao et al., 2017). The other example of workflow on mineral predictivity is on 3D modeling by Li et al (2019) reproduced in Fig. 3.4.

3.2.1.1 Limitation of mineral predictivity mapping

Mineral prospectivity mapping is a method more applicable at regional to camp scale (Porwal, Kreuzer, 2010). It is less effective and less frequently used at the project to deposit scale where it competes with direct detection techniques (McCuaig and Hronsky, 2000; Hronsky and Groves, 2008; McCuaig et al., 2010; Porwal and Kreuzer, 2010). Knowledge-driven and data-driven approaches are generally considered mutually exclusive, thus running each of the approach separately is necessary but in so doing, some available geological information remains underutilized in both approaches. However, hybrids such as fuzzy weights-of-evidence and neuro-fuzzy models have been developed to bridge this gap and optimize utilization of both conceptual and empirical (e.g., mineral deposits) data in mineral prospectivity mapping (Porwal and Kreuzer, 2010; Li et al., 2019).

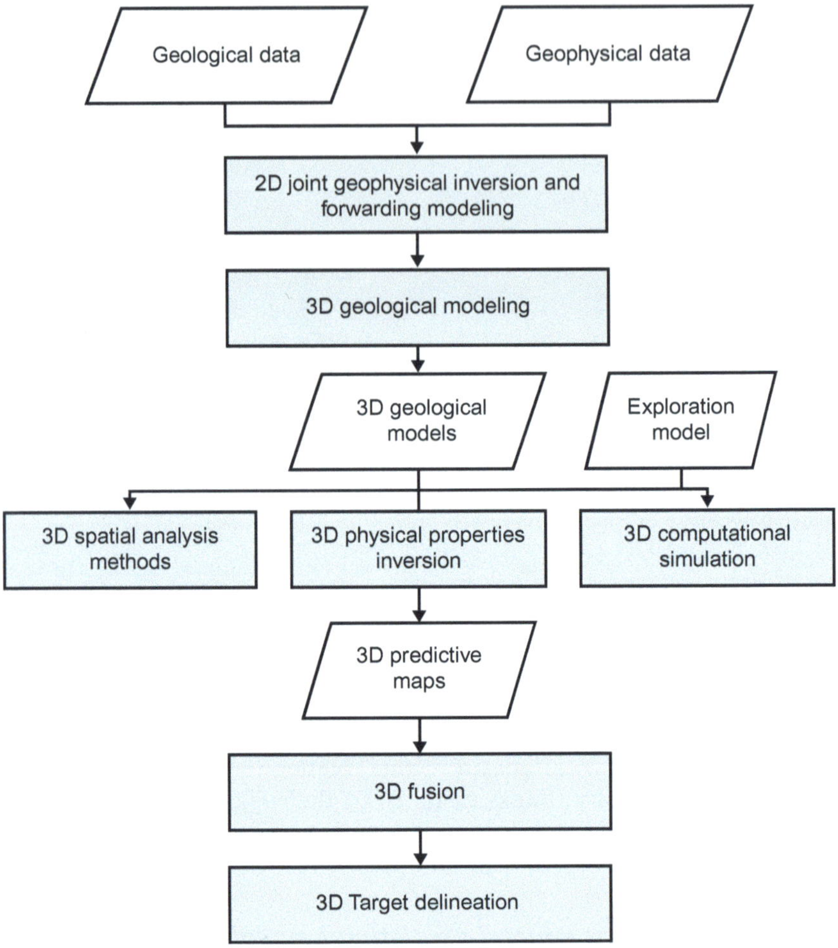

FIGURE 3.4 Flow chart on mineral predictivity used by Li et al. (2019).

3.2.2 Lithogeochemistry

The application of lithogeochemistry in recent years has increased particularly in areas where ore signatures are wholly or nearly wholly screened off from the surface due to lateritization or covered by thick overburden such that by applying conventional techniques of establishing geochemical anomaly(ies) in either soil or stream samples, the geochemical haloes/signatures are too weak to be recognized on the surface. Similarly, highly weathered profiles are generally highly leached and depending on the degree of leaching, some elements are leached down profiles and an attempt to undertake soil or stream sampling in such areas turns to be unsuccessful. Contrarily, for some elements such as Ni, laterites act as accumulation agents such that any small amount of Ni available on the surface over a barren area may be accumulated beyond background values as a result of pedogenic processes (e.g., Brand, 1999; Macheyeki, 2011). Therefore, understanding both the type of materials being sampled and the underlying bedrock will help to ascertain whether an area is worth exploring for a given metal (mineral) or not-hence, lithogeochemistry.

Lithogeochemistry is a compound word for "lithos-" and "geochemistry" which literally refer to "rock units-" and "the chemistry of rocks." Lithogeochemistry refers to those studies that involve geochemical exploration studies in the primary environment (Govett, 1983; Cohen and Bowell, 2014). In other words, lithogeochemistry refers to the use of geochemical signatures (elemental values, ratios between pairs of elements or a set of) to locate ore deposits in a certain geological environment (e.g., Govett, 1978; Macheyeki, 2011). Such an application of geochemical signatures is sometimes referred to as "lithogeochemical vectoring" toward ore environments. Attempts to locate Ni–Cu deposits using lithogeochemical vectors in different Archean to early Proterozoic geological environments, have been performed by a number of workers. Brand (1999) presents a trace element Ni–Cr ratio. He reports that Ni–Cr ratios >1 corresponds to potentially mineralized komatiite flows which have high Ni and low Cr as opposed to barren ones, which have high Cr and low Ni. The environments with potential to host Ni-sulfides have relatively high MgO, Ni, and Ni:Cr ratios and low Cr concentrations. This reflects a primary mineralogy dominated by olivine and alteration products. On the other hand, Cr content increases in non-ore environments along with MgO, Ni, and Ni:Cr ratios decreasing away from the potentially channelized mineralized environments (Brand, 1999). In the regolith environment, the absolute content of Ni effectively increases Ni:Cr ratios (>1) due to association of Ni with Mg-bearing minerals and Fe–oxides (Smith, 1984; Brand, 1999). On the other hand, the Cu:Zn ratio can assist in discriminating Ni-sulfide mineralized zones from barren ones because Ni sulfides have greater concentrations of Cu than Zn, thus the combined $Ni/Cr \times Cu/Zn$ vector is reported by Brand (1999) to have been successfully used in locating ores in basal komatiites.

Yinggui et al. (1995) report that $Cr \times Ag > 30$ indicate dimensions of Ni-Cu ore bodies. $Cr \times Ag > 100$ showing cutoff grades, whereas $Cr \times Ag > 300$ being capable of defining economic Ni–Cu ore bodies in mafic and ultramafic bodies in the Karatüngk Ni–Cu deposit in NW China. Two nickeliferous mafic intrusives in this area were intersected by completion of their study. Although their geochemical vector expressions were relatively complex, as different levels of geological profiles were assigned to different ranges of element ratios, the use of geochemical techniques seems to be effective in ore target generation. Similarly, Macheyeki (2011) reports that for the Kabanga type Ni–Cu sulfides (and probably for the search for Ni–Cu sulfides in the same geological environment), (Pd/V) × 1000 and [(Pd/V)/(Cu/Cr)] × 100 are useful vectors or element ratios provided that attention is paid more on the patterns of ratios than on the absolute values of the same.

3.2.2.1 Principal component analysis and lithogeochemical vectors

In order to obtain lithogeochemical vectors, one has to obtain element associations of ores, host rocks, and country rocks. The best method is the principal component analysis (PCA). The concept and method of PCA were first introduced by Pearson (1901), and further developed by Hotelling (1933).

The PCA is a technique for reducing large number of variables to a few uncorrelated variables, so that a variation diagram may contain information about a large number of variables. The PCA is used in order to establish different associations among variables or elements in a large area in order to minimize bias in geochemical data. Transformation of the data is undertaken with no statistical assumptions and that always PCA is a variance-oriented that shows the total structure in the data (all variables are "forced" into the result) and accounts for maximum variance of all observed variables. The PCA transforms a number of potentially correlated variables (here: the elemental abundances) into a smaller number of uncorrelated variables: the "principal components", whose covariance matrix is diagonal (d'Uston and Gellert, 2006), that is, it results in orthogonal (uncorrelated) factors. The PCA therefore removes the variance and seeks a second linear combination which explains the maximum proportion of the remaining variance, and so on (Atchley, 2007). Principal components (from PCA) reflect both common and unique variance of the variables and may be seen as a variance-focused approach seeking to reproduce both the total variables with all components and to reproduce the correlations, so-called Eigenvectors. Eigenvectors are principal component (PC) coefficients or loadings that describe the relative significance of a component (or chemical element) and its variability within the data set and are used to automatically calculate scores for each principal component. The element loading values determine a sample point's score, so that grouping of high loading elements give high scores and grouping of low loading elements give low scores. The amount of eigenvectors calculated will be equal to the numbers of variables used (Reimann et al., 2008), which in geochemical surveys is the number of chemical elements (Dempster et al., 2013). Eigenvectors should not be confused with Eigen values. From the Scree Plots (Fig. 3.5), an eigenvalue is the measure of PC contribution relative to other PCs in a multivariate data set.

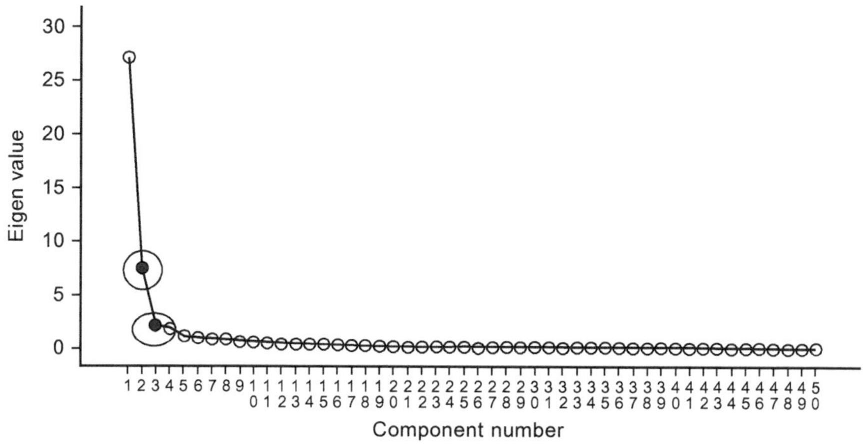

FIGURE 3.5 Scree plot for a data set. Components to the right of component 5 (below Eigen value of 1.0) do not account for a significant portion of variance in the data set. Source: *From Zakaria, N., 2014. Body shape analysis and identification of key dimensions for apparel sizing systems. Anthropomet., Apparel Sizing Design, 95–119. doi:10.1533/9780857096890.1.95.*

It is recommended that before applying PCA softwares, one has to make sure that the data is verified and the possible number of factors needed to explain for the pattern of relationships among variables is established. Further, the relationship of variables with the geology has to be worked out (for instance by plotting all raw data on the geology map), and proportion of the elements explained by the most important PCs and the possible spatial relationship of the factors estimated. It is also important to undertake log-ratio transformation, or else results may be uninterpretable (Chen et al., 2015). The three widely used log–ratio transformations are:

i. the additive log-ratio transformation defined by

$$y_i = \log\left(x/x_D\right)\ (i = 1, \ldots, D\text{-}1) \tag{3.3}$$

where x_D is a compositional component of choice,

ii. the centered log–ratio transformation (Aitchison, 1986) defined by

$$z_i = \log\left(x_i/g(x_D)\right)\ (i = 1, \ldots, D) \tag{3.4}$$

where $g(x_D)$ is a geometric mean of composition and

iii. the isometric log-ratio transformation (Egozcue et al., 2003) which represents combinations of elements representing balances that result in an orthonormal space:

$$y1 = \sqrt{\mathrm{i}/(\mathrm{i}+1)}\ln\frac{\mathrm{g}(\mathrm{x}1, \ldots, \mathrm{xi})}{(\mathrm{x}1-1)}(\mathrm{i} = 1, \ldots, \mathrm{D}-1) \tag{3.5}$$

The application of a centered log–ratio transformation will provide more reliable and sound results. The isometric log–ratio method may be used for data sets where balances between the elements are constructed and provides orthonormal basis in the compositional data space (Chen et al., 2015).

Once the data have been verified as stated above, the next step would be to normalize the data using either logarithmic, square root or Box–Cox transformations. In logarithmic transformation, natural logarithms of data values are used to transform the data set, generally log(x + 1) in order to prevent the occurrence of negative results for values less than 1 (Sara, 2011).

The square root transformation involves taking the square root of data values. This transformation method is used when the variable is a count and Box–Cox transformation is a power transformation type defined by:

$$y_i^{(\pi)} = \begin{cases} \dfrac{(y_i^{(\pi)} - 1)}{\lambda} & \text{when } \lambda \neq 0 \\ \log(y_i) & \text{when } \lambda = 0 \end{cases} \tag{3.6}$$

where λ is a parameter based on maximum likelihood estimation and y is the function.

At this stage, sample skewness is computed to assess whether the data set fits a normal distribution, having positive or negative skewness or not (Fig. 3.6A and B). Autoscaling (also called z-transformation) and Pareto scaling, as scaling factor, can be used (Sara, 2011).

In case outliers are obtained in the process of data normalization, they should be removed using sophisticated techniques such as Mahalanobis distance calculation (Drouin, 2012) because, if outliers are not removed, they can affect the data negatively. This pre-treatment procedure leads to identification of clusters as well; clusters are literally groups within the data set.

The next stage is to identify number of PCs. This can be achieved by using scree plots (Fig. 3.6). With scree plots or as also called *Kaiser criterion*, only PCs with eigenvalue (variance) greater than 1 are retained—these PCs usually represent 80%–90% of cumulative proportion of variance (Sara, 2011).

Once the number of PCs has been obtained, they are then multiplied by their component coefficients or as called "loadings." Component coefficients describe the relative significance of a component (or chemical element) and its variability within the data set. They are the correlation coefficients between the variables (rows) and factors (columns), analogous to Pearson's correlation coefficient "r" (Fig. 3.7).

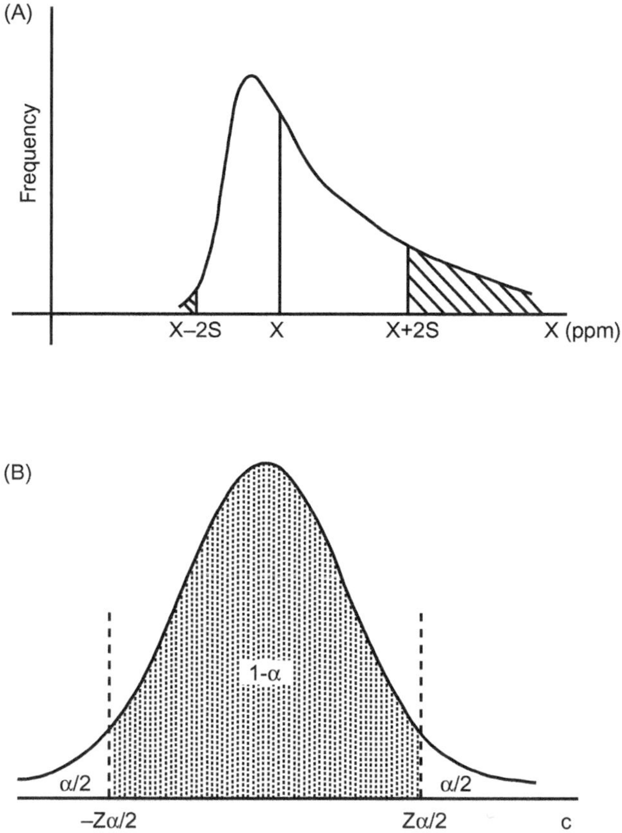

FIGURE 3.6 Diagram illustrating data distribution. (A) Raw data, positively skewed distribution. X = average background value: S = standard deviation (Shi and Wang, 1995). (B) Example of a normal distribution data plot. Source: *From Pounis, G., 2019. Statistical analysis of retrospective health and nutrition data. Anal. Nutr. Res., 103–144. doi:10.1016/b978-0-12-814556-2.00005-1.*

To get the percent of variance in all the variables accounted for by each factor, add the sum of the squared factor loadings for that factor (column) and divide by the number of variables (rows). The reader has to note that the number of variables equals the sum of their variances as the variance of a standardized variable is 1. This is the same as dividing the factor's Eigen value by the number of variables.

The next stage is to compute for *score values*. This is achieved by multiplying PCs with component coefficients. The score values are then plotted on the Score plot (Fig. 3.8A and B).

Upon plotting of the score values in the score plot, one can clearly see element groupings that reflect important information. In the case of

	Ca	Cu	Fe	K	Mg	Na	P	Sr
Cu	■ 0.25 ⊙ 0.84 + 0.77 ▲ 0.07							
Fe	■ 0.15 ⊙ 0.92 + 0.79 ▲ -0.06	■ 0.59 ⊙ 0.91 + 0.76 ▲ 0.7						
K	■ 0.37 ⊙ 0.51 + 0.77 ▲ 0.02	■ 0.51 ⊙ 0.18 + 0.76 ▲ 0.5	■ 0.43 ⊙ 0.28 + 0.9 ▲ 0.38					
Mg	■ 0.32 ⊙ 0.96 + 0.57 ▲ 0.41	■ 0.75 ⊙ 0.83 + 0.73 ▲ 0.82	■ 0.58 ⊙ 0.91 + 0.86 ▲ 0.66	■ 0.53 ⊙ 0.56 + 0.87 ▲ 0.56				
Na	■ 0.9 ⊙ -0.35 + 0.57 ▲ 0.42	■ 0.32 ⊙ -0.58 + 0.36 ▲ 0.65	■ 0.08 ⊙ -0.56 + 0.34 ▲ 0.2	■ 0.43 ⊙ 0.52 + 0.36 ▲ 0.56	■ 0.4 ⊙ -0.35 + 0.19 ▲ 0.71			
P	■ 0.76 ⊙ 0.28 + 0.7 ▲ 0	■ 0.13 ⊙ 0.13 + 0.59 ▲ 0.5	■ 0.37 ⊙ 0.29 + 0.43 ▲ 0.66	■ 0.42 ⊙ -0.19 + 0.41 ▲ 0.5	■ 0.14 ⊙ 0.03 + 0.36 ▲ 0.39	■ 0.57 ⊙ -0.23 + 0.21 ▲ 0.17		
Sr	■ 0.75 ⊙ -0.4 + 0.22 ▲ 0.4	■ 0.36 ⊙ -0.32 + 0.14 ▲ 0.28	■ 0.11 ⊙ -0.51 + 0.03 ▲ 0.2	■ 0.39 ⊙ 0.14 + 0.1 ▲ 0.19	■ 0.32 ⊙ -0.36 + -0.01 ▲ 0.46	■ 0.82 ⊙ 0.65 + 0.8 ▲ 0.45	■ 0.5 ⊙ 0.44 + -0.08 ▲ 0.23	
Zn	■ 0.26 ⊙ -0.08 + 0.91 ▲ -0.11	■ 0.74 ⊙ 0.28 + 0.75 ▲ 0.66	■ 0.62 ⊙ -0.01 + 0.81 ▲ 0.83	■ 0.52 ⊙ -0.16 + 0.83 ▲ 0.4	■ 0.61 ⊙ -0.02 + 0.58 ▲ 0.52	■ 0.3 ⊙ 0.07 + 0.61 ▲ 0.11	■ 0.29 ⊙ -0.39 + 0.44 ▲ 0.65	■ 0.48 ⊙ 0.67 + 0.36 ▲ 0.1

■ GRANITES
⊙ ALKALINE ROCKS
+ BASALTS
▲ SEDIMENTS

FIGURE 3.7 Pearson correlation coefficients within four selected rock units underlying the C-horizon in the survey area for nine elements (Granites: $n = 94$, Alkaline rocks: $n = 12$, Basalts: $n = 42$, sediments: $n = 68$). Note that the same pair of elements can show positive as well as negative correlation depending on rock type (Reimann et al., 2002).

Figure 2.8(A), four groups are identified from the score plot (1) Alikali-related elements, (2) biogenetic-related elements, (3) Fe–oxyhydroxides, and finally (4) heavy minerals–related elements. Numbers (1)–(4) represent groups of elements that can be used for different purposes including computation of element ratios as shown in Section 2.3.2.2 for the Kabanga Nickel project.

Pereira et al. (2003) use the PCA to identify anomaly candidates in a geochemical data set. The underlying pre-requisite is only that the samples are representative of background/anomaly multivariate structure (in the sense that it allows for choosing "anomaly candidates" on the PCA of Standardized Data graphs).

Further to the outstanding power of PCA in lithogeochemistry, PCA in combination with geophysics, GIS and remote sensing can help to locate potential mineral deposit targets both near surface and Burren ore bodies (Ott et al., 2006) (Fig. 3.9).

The reader is referred to Le Maitre (1982), Aitchison (1986), Jimenez-Espinosa et al. (1993), Rollinson (1993), Kline (1994), Jolliffe (2002),

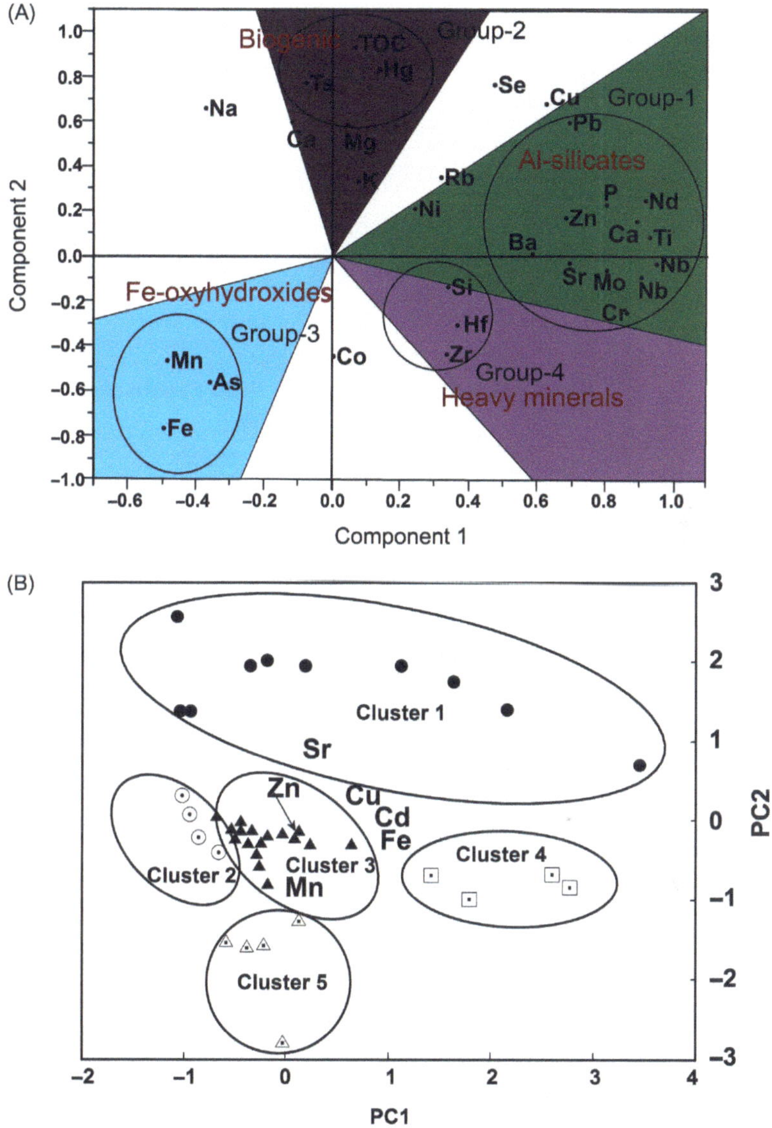

FIGURE 3.8 (A) Score plot, principal component I (PC1) versus principal component II (PC2) for geochemical elements from Serra dos Carajàs, Southeastern Amazon, Brazil Brazil (Sahoo et al., 2016). (B) Bivariate plot (Score plot) of the first and second principal components scores from the PCA of the inorganic geochemical data from the PFB catena soils. The variable coordinates for Cd, Cu, Fe, Mn, Sr, and Zn are plotted using elemental symbols. The five ellipses encircle groupings identified by cluster analysis. Clusters contain data of samples from various places, NW Belize. Source: *From Beach, T., Ulmer, A., Cook, D., Brennan, M.L., Luzzadder-Beach, S., Doyle, C., et al., 2018. Geoarchaeology and tropical forest soil catenas of northwestern Belize. Quatern. Int. 463, 198-217. doi:10.1016/j.quaint.2017.02.031.*

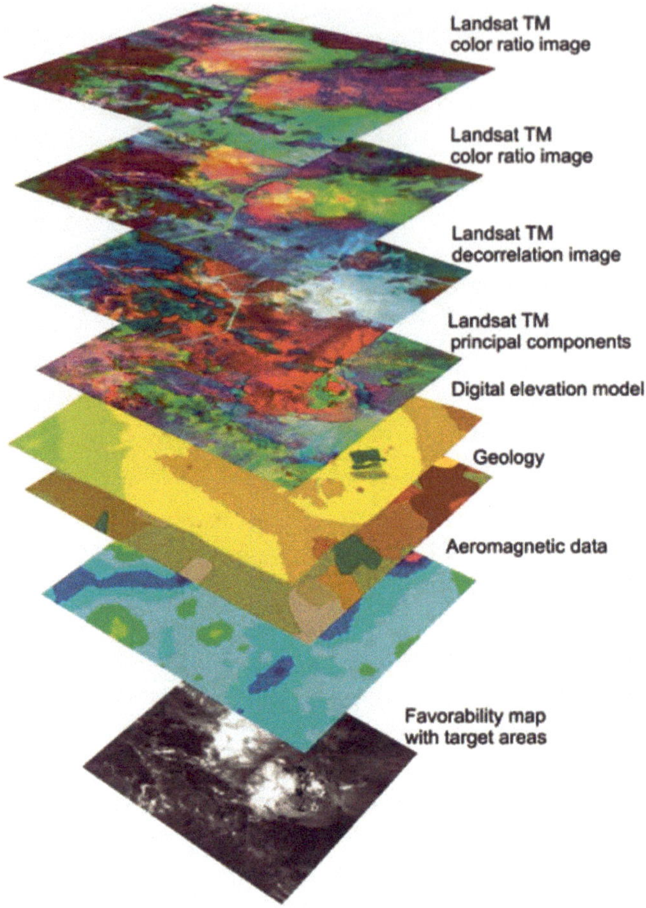

FIGURE 3.9 Perspective view of different thematic layers of the database in the vicinity of La Escondida mining district. Upper layers represent optimized Landsat data derived from band rationing, principal component analysis (PCA), and inverse PCA. Lower layers represent topographic data, lithology, and aeromagnetic data. Bottom layer is one of the calculated favorability maps (Ott et al., 2006).

Egozcue et al. (2003), d'Uston and Gellert (2006), Atchley (2007), Tabachnick and Fidell (2007), Reimann et al (2008), Field (2009), Dempster et al. (2013), and Chen et al. (2015), for details.

3.2.2.2 *Limitation of lithogeochemistry*

Lithogeochemical vectors seem to be geology sensitive (Brand, 1999), ore-type sensitive (Muecke and Chatterjee, 1983; Rugless, 1983; Brand, 1999; Hosseini-Dinani and Aftabi, 2016) and although quantifiable lithogeochemical vectors are more appealing to use in mineral exploration, in practice, their patterns or trends are more important than their values (Macheyeki, 2011; Hosseini-Dinani and Aftabi, 2016). In other words, attention should be devoted more on the trend than the absolute values (Macheyeki, 2003, 2011).

3.2.3 Mineral zoning

Mineral zoning, an indication of metal deposition in relative order during primary crystallization or sedimentation. It occurs in a multiple series of hydrothermal depositional sources within a single ore body and between ore bodies (Haldar, 2018). Mineral zoning is a typical characteristic of epithermal deposits (e.g., White and Hedenquist, 1990), volcanomassive sulfides (VMS), skarn deposits, etc. Meinert Most large skarn deposits are zoned in both space and time relative to associated intrusions. As per (1997), zonation of minerals of a given deposit can occur from micrometers to kilometers and reflects infiltrative fluid flow, wall rock reaction, temperature variations, and fluid mixing. Citing skarn deposits, Meinert (1997) shows a general zonation pattern that occurs in these deposits: proximal garnet, distal pyroxene and vesuvianite (or a pyroxenoid such as wollastonite, bustamite, or rhodonite) at the marble front.

In W-skarns, garnet is commonly sub-calcic and the pyroxene is Fe-rich reflecting particularly reducing wall rocks or great depth of formation. Meinert (1997) also shows that Sn-skarns, can have sub-calcic garnet and Fe-rich pyroxene, but this reduced mineral assemblage is typically due to an association with reduced S-type granites. The Sn-skarns differ from most other skarn types in having a late greisens stage that may replace earlier Sn-bearing calc-silicate minerals, thus liberating Sn to form cassiterite.

High-grade Au-skarns have low ratios of garnet:pyroxene and are associated with both reduced plutons and reduced wall rocks. Similarly, Zn–Pb-skarns tend to have low ratios of garnet:pyroxene and generally form distal to associated intrusions. Differences in oxidation states correlate well with different skarn zonation patterns particularly garnet: pyroxene ratios and compositions and can be used in both classification and exploration for skarn deposits. Mineral or element zoning can also be viewed in the context of the ore itself. The study of the systematics of internal ore-deposit variation is the study of zoning, defined as the spatial distribution patterns of major or trace elements, mineral species, mineral assemblages or textures in ore deposits. Ore deposit zoning can extend from vein-let, to vein systems, to single ore shoots and to regional scale zoning.

Mineral assemblages can also be zoned, minerals as individuals can also be found to be zoning and so are zoning due to mineral phases (silicate phase, sulfide phase, etc.). Zoning within the ore can also take the form of minor chemical changes that are invisible to the eye, so-called cryptic zoning or chemical zoning or it may take the form of changes in the texture of a rock or vein (textural zoning). Overall zoning deals with arrays of ore deposit components and characteristics in three dimensions (www.unalmed.edu.co/~rrodriguez/Ore-Genesis-Notes/Hydrothermal%20Alteration%20Systematics.htm).

The distribution of metals and sulfide types is commonly zoned on the scale of an individual lens and in clusters of lenses. Copper for example, is usually high relative to Zn + Pb in the core of ore pipes and in the spines of massive sulfides (e.g., Wilkinson, 2014). Therefore, the ratio of Zn + Pb: Cu increases around the outside of the pipe and towards the upper part and margins of the massive zone. Gold and Ag usually are highest in the fringe areas and so is Ba. The proportions of Zn, Pb, and Ba also tend to increase in lenses peripheral to the center of the deposits, both laterally and vertically (up-stratigraphy). Pyrrhotite and magnetite may occur in the core zone with pyrite usually becoming dominant at the fringes (e.g., Ames et al., 1993).

Quantified zonation models can be used to predict for blind mineral targets (White and Hedenquist, 1990; Parsapoor et al., 2017). White and Hendenquist (1990) point out that alteration zoning can be used as a pointer toward the most prospective part of Au systems and other mineral systems but only when the styles of zonations have been correctly recognized (Fig. 3.10). Fig. 3.10 shows a regional mineral zonation that when well studied, it saves as a model for successive studies in the area

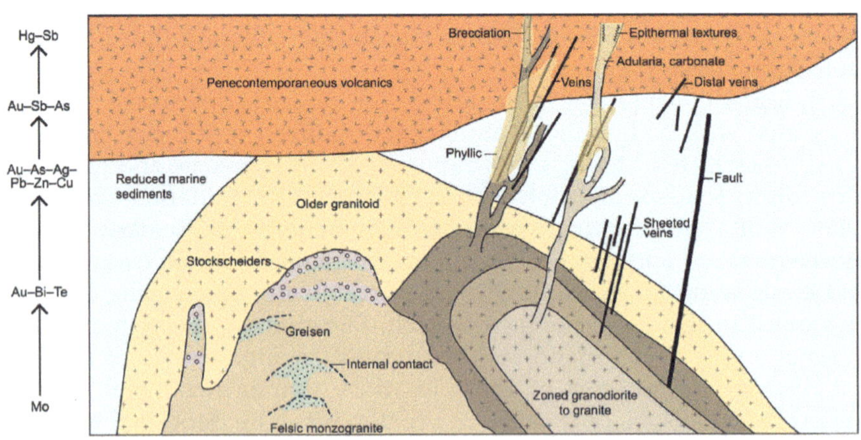

FIGURE 3.10 Geological model of intrusion-related Au systems identified in the Southern New England Orogen. Vertical extent of zonation shown in the model approximates a 3–5 km crustal section from plutonic to typical porphyry emplacement levels (Ford et al., 2019).

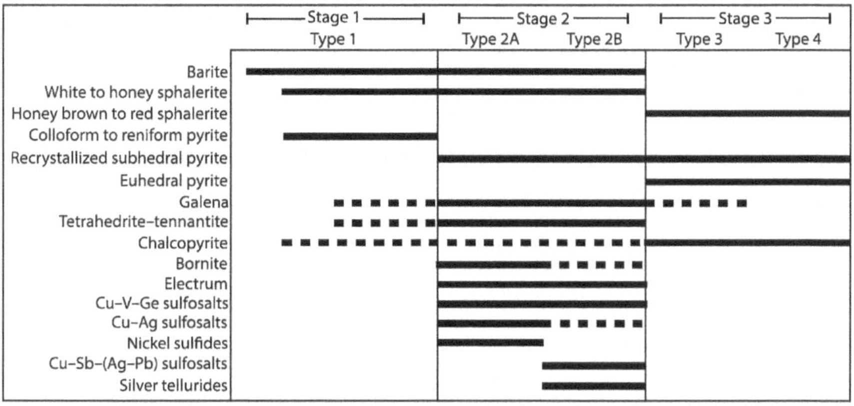

FIGURE 3.11 Simplified paragenetic scheme highlighting the three main stages of mineral deposition at the Zn–Pb–Cu–Ag–Au Lemarchant volcanogenic massive sulfide (VMS) deposit, Newfoundland, Canada (Gill et al., 2019).

or in areas with similar geological settings. Fig. 3.11 shows a simplified paragenetic scheme of the three main stages of mineral deposition at the Zn–Pb–Cu–Ag–Au Lemarchant volcanogenic massive sulfide deposit.

3.2.3.1 The role of Factor Analysis in obtaining mineral zonations

Factor analysis can be effectively used to study symbiotic combination of elements and can be used to realize division of different geochemical backgrounds (Shi et al., 2004; Liu et al., 2015; Xiang et al., 2018). Further, it can assist in reflecting regional metallogenic geological background and it can also assist in revealing metallogenic element spatial variation characteristics of the metallogenic element and its elemental combination (Xiang et al., 2018).

In order to obtain mineral zonation one needs to establish the element groups from the data set using either Factor analysis or PCA. Suppose from a data set of a blind ore body, you have elements Ag, As, Ba, Be, Bi, Ca, Cd, Ce, Co, Cr, Cs, Cu, Fe, and Ga. Suppose also that after Factor analysis, the groups defined by (1) Ag, (2) Ag–Ba–Be, (3) Ca–Cd, (4) Co–Cr–Fe are obtained. The four groups (1)–(4) above characterize the main element associations within the data set and reflect ore and gangue element/mineral variations within and outside the deposit. The next stage will be to establish the patterns of the element zonations. This can be achieved through many ways, one of which is (1) to enter the values of these elements/groups of each sample as columns (if samples are arranged in columns) with their values entered along rows. By plotting the groups taking into consideration of their coordinates, a 2D view can be obtained that can enable one to see the pattern of the elements/groups much easier. Ideally and in plan-view,

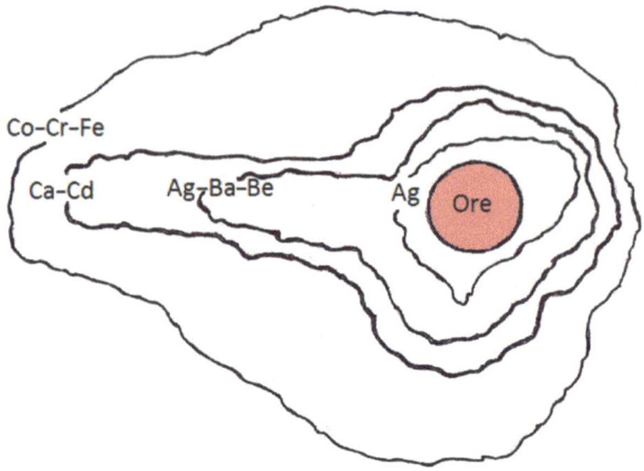

FIGURE 3.12 A conceptual 2D view of the elements/groups of elements as established from geostatistics (e.g., Factor analysis).

the plot that is obtained is like a contour map (Fig. 3.12). The 3D plots of other types of ore deposits can be obtained by taking into consideration not only the location and value of the sample, but also the elevation of the point where the sample was collected.

Factor analysis is a statistical method that is capable of reducing many variables into just few variables that we call factors. One may find a little difference between Factor analysis and PCA as explained in the previous section. While in PCA transformation of the data is undertaken with no statistical assumptions, Factor analysis assumes a statistical model and unlike PCA, true Factor analysis is correlation-oriented (Reyment and Jöreskog, 1996).

In Factor analysis, the emphasis is on joint variations in response to unobserved latent variables (i.e. variables that are not directly observed but are rather inferred through a mathematical model) from other variables that are observed (directly measured). The term "Factor analysis" therefore refers to a set of closely related models intended for exploring or establishing correlational structure among observed random variables (Basilevsky, 1994). If we have, say 20 observed random variations shown by a set of variables, it is possible that there could be only 3, 4, 5 unobserved or underlying variables that define the whole set of both observed and unobserved variables.

Key aspects that are considered for Factor analysis include geometrical interpretation of data matrices which requires basics of statistics. Each variable is expressed as a linear combination of factors. The factors are some common factors plus a unique factor. The factor model is represented as:

$$X_i = A_{i\,1}F_1 + A_{i\,2}F_2 + A_{i\,3}F_3 + \ldots + A_{im}F_m + V_iU_i \quad (3.7)$$

where

X_i = ith standardized variable
A_{ij} = standardized multivariate regression coefficient of variables i on common factor j
F_j = common factor j
V_i = standardized regression coefficient of variable i on unique factor i
U_i = the unique factor for variable i
m = number of common factors

The first set of weights (factor score coefficients) are chosen so that the first factor explains the largest portion of the total variance and thereafter a second set of weights can be selected, so that the second factor explains most of the residual variance, subject to being uncorrelated with the first factor. This same principle applies for selecting additional weights of additional factors.

The common factors themselves can be expressed as linear combinations of the observed variables:

$$F_i = W_{i1}X_1 + W_{i2}X_2 + W_{i3}X_3 + \ldots + W_{ik}X_k \tag{3.8}$$

where:

F_i = estimate of i th factor,
W_i = weight or factor score coefficient and
k = number of variables.

In order to undertake sound Factor analysis, one has to make sure that the problem at hand is well understood in terms of objectives, variables specifications and on variables measurements on an interval or ratio scale. Furthermore, the sample size has to be within acceptable range, that is, at least four or five times of the number of observations/sample size as there are variables.

Thereafter, correlation matrix should be constructed in order to be able to analyze correlation matrix between variables; for example if by using the Bartlett's test of sphericity and no data rejection is observed, then, the Factor analysis is not appropriate there (Maxwell and Delaney, 2000); and if the Kaiser–Meyer–Olkin (KMO) measure of sampling adequacy is small, then the correlations between pairs of variables cannot be explained by other variables and Factor analysis may not be appropriate.

The next stage would be to determine the method of Factor analysis to be employed followed by determination of the number of factors from the data set such as the use of Priori Determination. In using Eigenvalues (as is the case for PCA), only factors with Eigenvalues greater than 1.0 are retained.

The next stage would be to rotate the factors. Ideally, each factor should have significant loadings with only a few factors, if possible with only one. The rotation is called orthogonal rotation (Fig. 3.13) if the axes are maintained at right angles we can then use orthogonal rotation if we believe that data variance is not highly correlated to one another. For orthogonal rotation, selecting the number of factors to extract is difficult; however, the Scree plot provides some useful help (Reimann et al., 2002).

It is called Varimax procedure if the axes maintained are at right angles. It is an acceptable fact that orthogonal rotation results in uncorrelated factors. Further, the rotation is called Oblique rotation when axes are not maintained at right angles. Oblique rotation should be used when factors in the population are likely to be strongly correlated. As a general guide, rotated factors that have one or two variables should be interpreted with caution.

A factor can be interpreted in terms of the variables that load high on it. When variables are plotted using the factor loadings as coordinates, then, variables at the end of an axis are those that have high loadings on that factor, and hence describe the factor.

After establishing the behavior of variables as described above, the next step would be to calculate Factor Scores in an attempt to reduce the data so that they can be manageable (e.g., Basilevsky, 1994). In this process, descriptive methods can tell us groupings that appear to be

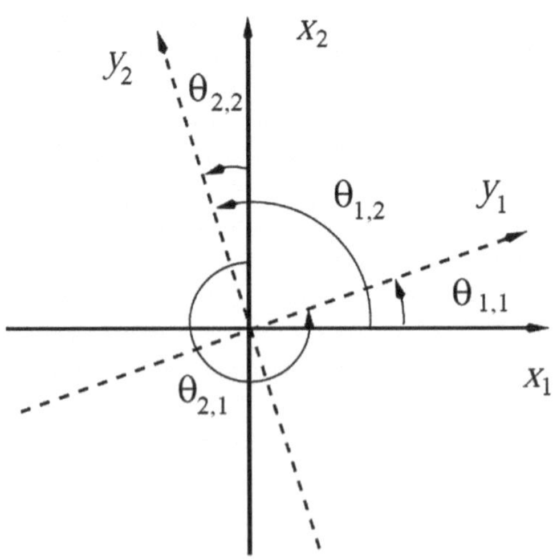

FIGURE 3.13 An orthogonal rotation in two dimensions. The angle of rotation between an old axis m and a new axis n is denoted by $\theta_{m,\,n}$ (www.utdallas.edu/»herve) (In: Lewis-Beck M., Bryman, A., Futing T. (Eds.) (2003). Encyclopedia of Social Sciences Research Methods. Thousand Oaks (CA): Sage.).

giving the same constructs (underlying correlation). In data reduction, we use descriptive analysis in order to reduce matrixes within the large data (e.g., Lebart et al., 1984). More common factors can then be extracted. Data extraction (usually extraction of eigenvalues or vectors) is done not only to extract more common data but actually to be able to retain the essential information content (e.g., Basilevsky, 1994). Lebart et al. (1984) have called this a log contrast PCA. In the course of data reduction and when employing data or factor scores computation using well-defined estimators, the choice is based on the particular factor model used (Basilevsky, 1994). The factor scores for the *i*th factor may be estimated as:

$$F_i = W_{i1}\, X_1 + W_{i2}\, X_2 + W_{i3}\, X_3 + \ldots + W_{ik}\, X_k \tag{3.9}$$

A factor score can be defined as a composite measure created for each observation on each factor extracted in the Factor analysis. The factor weights are used in conjunction with the original variable values to calculate each observation's score. The factor scores are standardized according to a *z*-score. Factor scores can be construed as the "concentration" of the new factors (Jöreskog et al., 1976) in each sample and are calculated by multiplying the concentration values of all the samples, including outliers that are removed during the data exploration process, by the factor score coefficients obtained from the Factor analysis (Drouin, 2012).

Factor scores or "factor loadings" indicate how each "hidden" factor is associated with the "observable" variables used in the analysis. Say you have five observable variables that identify two hidden factors. For example, suppose you have factor loadings on hidden Factor 1 across the five variables: 0.90, −0.35, 0.76, −0.89, 0.56. The first and third-factor loadings (0.90, 0.76) indicate that observable measures 1 and 3 can be used to "describe" hidden Factor 1; in other words, Factor 1 has characteristics very similar to what observable measures 1 and 3 measure. Observable variables 2 and 5 (−0.35, 0.56) are not useful to describe hidden Factor 1 because their factor loadings on hidden Factor 1 are too small (not > or = to 0.70). Finally, the negative factor loading of observable measure 4 with hidden Factor 1 (−0.89) means that hidden Factor 1 has the characteristic "opposite" of whatever observable measure 4 measures.

If the observable measures captured the dimensions "fun," "hard," "happy," "tough," and "sad," then hidden Factor 1 can be described as "fun" (loading = 0.90), "happy" (loading = 0.75), and "NOT Sad" (loading = −0.89), (www.researchgate.net)

Finally, a Model Fit should be worked out. The correlations between the variables can be deduced from the estimated correlations between the variables and the factors. The differences between the observed correlations (in the input correlation matrix) and the reproduced correlations (estimated from the factor matrix) can be examined to determine model fit. These differences are called *residuals*.

A determining factor in Factor analysis is based on the assumption that there is a linear relationship between the factors and the variables when computing for the correlations (Gorsuch, 1983). Further, there should be ≥3 variables, although this depends on the design of the study (Tabachnick and Fidell, 2007). A factor with two variables is reliable when the variables are highly correlated with each other ($r^2 > 0.70$) but fairly uncorrelated with other variables. The recommended sample size is ≥300, and the variables that are subjected to Factor analysis should have ≥5–10 observations (Comrey and Lee, 1992). Guadagnoli and Velicer (1988) proposed that if the dataset has several high factor loading scores ($r^2 > 0.80$), then a smaller small size ($n > 150$) should be sufficient. A factor loading for a variable is a measure of how much the variable contributes to the factor; thus, high factor loading scores indicate that the dimensions of the factors are better accounted for by the variables. Next, the correlation r^2 must be ≥0.30 since anything lower would suggest a really weak relationship between the variables (Tabachnick and Fidell, 2007).

In an ideal or simplified way, let's assume we have geochemical data arranged in 20 columns (U, Ag, As, Co, Cr, Cs, Cu, Hf, Li, Lu, Mn, Mo, etc.) and each element has been analyzed in 16 samples arranged in rows. Let us also consider that those samples and their resulting set of elements were obtained from analysis of soil samples that were taken over both igneous and sedimentary rocks (Table 3.2). We therefore have over 300 data to work with. Our key task would be to see if such elements (variables) are correlated to each other or not. From that, we can extract the underlying constructs (factors).

Suppose also that, after running the Factor analysis, the results show elements Ni, Cu, Cr, and Co as highly correlated ($r^2 = 0.9$) and elements V, U, and Li as well correlated ($r^2 = 0.8$). Let us also assume that the elements Rb and As are also showing fair correlation ($r^2 = 0.5$). It will therefore imply that Ni–Cu–Cr–Co are a one factor, V–U–Li are another factor and Rb–As are a factor. Therefore, it is possible to infer that Ni–Cu–Cr–Co are related to mafic or ultramafic rocks, V-U-Li to spodumene ($LiAl(SiO_3)_2$) mineralized pegmatite and Rb–As could be related to deep-seated fluids (in this case through fractures) from felsic rock sources. From such a factor analysis we reduce the variables to just three factors. Other elements that could have correlation less than 0.5 might actually not be related to each other and may therefore not be grouped together.

In geology or mineral exploration, the moment such a relationship (e.g., Ni–Cu–Cr–Co) in the data set is identified, it immediately tells

TABLE 3.2 Trace elements obtained from one region in Tanzania.

Sample ID	U	Ag	As	Co	Cr	Cs	Cu	Hf	Li	Lu	Mn	Mo	Nb	Ni	Pb	Rb	V	W	Zn	Zr
TZR1	2.26	0.3	11	7	22	1.4	8.4	1.9	3	0.13	360	0.8	19.4	12.3	20.1	90.5	37	0.6	14	41.4
TZR2	2.04	0	11	5	20	1.4	7.7	1.6	3	0.12	260	1.2	17	13.3	18.4	80	38	0.5	14	41.1
TZR3	2.45	0.3	12	5	36	1.5	6.9	1.7	4	0.15	220	1.2	14.8	10.6	20.5	76.9	51	0.6	16	45.4
TZR4	2.18	0.4	0	4	8	0.7	7.7	0.9	3	0.08	440	1.1	7.4	5	14.9	60.1	13	0.2	3	25.3
TZR5	2.96	0	9	3	14	1	6.4	1.1	3	0.14	220	0.6	9.7	4.8	15	70.7	19	0.3	7	27.3
TZR6	2.31	0	8	2	12	0.9	7.4	1.2	3	0.12	320	0.3	10.2	4.3	15.6	77.6	21	0.3	7	34.5
TZR7	2.08	0.3	8	2	11	0.6	9.1	1.3	3	0.12	380	0.5	14.3	3.6	12.7	54.1	20	0.2	4	42.1
TZR8	1.69	0.3	8	2	8	0.6	5.3	1.2	3	0.1	300	0.7	12.2	3.3	12.6	54.6	18	0.2	4	37
TZR9	2.13	0	10	6	12	1	7.3	1.1	2	0.11	660	1	15.6	5.3	19.9	73.4	20	0.3	6	34.4
TZR10	3.82	0.7	8	4	19	1.3	8.6	1.3	3	0.18	300	0.6	14.9	14	19	71.5	27	0.3	11	36.5
TZR11	3.25	0.4	10	3	14	1.4	10	1.4	3	0.2	220	0.3	13.3	15.8	19.4	78.4	28	0.3	9	36.4
TZR12	1.46	0.5	8	4	15	1	8.8	1.6	3	0.13	400	0.6	17.1	6.4	16.9	70.5	26	0.3	8	47.6
TZR13	2.00	1.0	8	5	5	2	8	3.2	1	0.10	300	1	12.6	4.3	20.9	60.4	22	0.2	5	30.1
TZR14	3.80	0.6	7	4	18	1.2	8.2	1.5	3	0.18	200	0.6	11.9	11	15	80.0	21	0.1	10	33.5
TZR15	3.27	0.3	9	3	13	1.2	9	1.2	3	0.2	120	0.3	9.3	14.3	15	69.1	19	0.2	8	36.0
TZR16	1.48	0.4	7	4	14	1	7.5	1.5	3	0.13	300	0.6	15.1	5.4	14	68.5	21	0.3	7	45.5

the explorer about the possible mafic/ultramafic source (be it volcanic or intrusive). It therefore narrows down our focus and improves the chances to get a corresponding ore body if it exists. If on another hand, a geologist finds an element relationship such as Ce, Mo, Ni, Zr, Nb, Sn, Th, Ca, Rb, and K, then, carbonatite source is likely (RSC, 2015). The mafic/ultramafic rocks, the pegmatites, and carbonatites that we infer them to be related to our geochemical data are actually the unobserved latent variables (i.e., variables that are not directly observed but are rather inferred through a mathematical model).

In the processing of Factor analysis, Univariate and multivariate normality within the data are also present (Child, 2006) and so are multivariate outliers. In the data set, as is for PCA, it is recommended to remove univariate and multivariate outliers (Field, 2009) and the different populations present in the data should be treated separately (Reimann et al., 2002) using sophisticated techniques like the calculation of Mahalanobis distance to identify multivariate outliers (Drouin, 2012).

3.2.3.2 Limitations of factor analysis and comparison with PCA

One of the limitations of this technique is that naming the factors can be problematic. Factor names may not accurately reflect the variables within the factor. Further, some variables are difficult to interpret because they may load onto more than one factor which is known as split loadings. These variables may correlate with each other to produce a factor despite having little underlying meaning for the factor (Tabachnick and Fidell, 2007). Finally, researchers need to conduct an orientation study using a large sample size at a specific point in time to ensure reliability for the factors. It is not recommended to pool results from several samples or from the same sample at different points in time as these methods may obscure the findings (Tabachnick and Fidell, 2007). As such, the findings from Factor analysis can be difficult to replicate.

Both PCA and Factor analysis are characterized by eigenvectors, eigenvalues, loadings, and scores. Both Factor analysis and PCA can be thought of as trying to represent some aspect of the covariance matrix Σ (or correlation matrix) as well as possible (Jolliffe, 2002).

However, main differences are, (1) PCA is essentially a purely mathematical operation in nature whereas PCA is statistical in nature; (2) while both PCA and factor analysis are multivariate based analyses, PCA is used as a simple starting point of the multivariate analysis; (3) PCA eigenvectors cumulatively account for all the variability in the dataset whereas Factor analysis results include an unresolved component; (4) Factor analysis results are often transformed through Varimax and other methods to optimize eigenvectors for interpretation. In summary, PCA concentrates on the diagonal elements, whereas Factor analysis is concerned about the off-diagonal elements (Jolliffe, 2002). As per the scope of this book, only basic principles on factor analysis have been given. The reader is referred to Guadagnoli and Velicer (1988), Tabachnick and Fidell (2007), Jolliffe (2002) and Yong and Pearce (2013) for details in factor analysis.

3.2.3.3 *Staged factor analysis*

The staged factor analysis, a multivariate statistical method, categorizes geochemical data and determines indicator elements related to target deposit (Ostadhosseini et al., 2018) and their elemental paragenesis (Yousefi et al., 2014; Yousefi, 2017; Fyzollahhi et al., 2018). The method operates similar to PCA but with a decreased correlation or covariance matrix. Using MATLAB software, noisy, or nonindicator elements are progressively removed in the data set until such a time when satisfactory significant multielement signature is obtained (Yousefi et al., 2014; Yousefi and Carranza, 2015).

According to Fyzollahhi et al. (2018), the first step is to apply PCA on the data in the classical way in order to extract common factors. The second stage is to rotate the data so that data with eigenvalues of >1 are retained for interpretation (e.g., Kaiser, 1958). From there threshold values are obtained for loadings in factor analysis in order to extract significant multielemental signature of the ore-type sought. By so doing, factors and their components with lower noises are separated (e.g., Davis, 2002; Yousefi et al., 2014; Fyzollahhi et al., 2018).

3.2.4 Concentration gradient modeling

Concentration gradient model is one of the most powerful exploration tools whereby concentration variations are used to tell about the characteristics of the geochemical anomaly. In this case, near-surface geochemical anomalies correspond with large concentration changes whereas less concentration variations are more likely to be related to the deeper geochemical anomalies (Ziaii et al., 2019). This technique is capable of distinguishing geochemical data related to mineralization and those related to barren zones without undertaking exploration drilling (Safari et al., 2018; Ziaii et al., 2019) and can be achieved through (1) analysis of element associations representing the supra-ore gradient and sub-ore gradient halos of mineral deposits; (2) analysis of a single component gradient, implying false anomaly; (3) analysis of mean values of indicator elements gradient outside significant geochemical anomalies to eliminate background noise in data analysis; and (4) mapping multiplicative geochemical gradient anomalies (Safari et al., 2018).

In this model, the gradient is calculated. This refers to the rate of element content change in unit distance—a value which is equal to the difference of content between each two adjacent samples (Safari et al., 2018; Ziaii et al., 2019). In order to obtain this gradient (Gr), the difference between element concentrations in two adjacent samples Δc (in ppm) is divided by Δx which is the distance (in m) in a given direction (x) between two adjacent samples (Sochevanov, 1961). By so doing the gradient expressed in ppm/m is obtained as:

$$Gr = \frac{\Delta c}{\Delta x} \tag{3.10}$$

In two dimensions, the gradient is given by Gonzalez and Woods (2008) as:

$$\nabla f = grad(f) = \begin{bmatrix} g_x \\ g_y \end{bmatrix} = \begin{bmatrix} \frac{g_y}{g_x} \\ \frac{g_f}{g_y} \end{bmatrix} \tag{3.11}$$

where ∇f is the vector with geometrical property pointing in the direction of greatest rate of change at location (x, y) and the formula $\partial f/\partial x$ would be the gradient in the x direction and $\partial f/\partial y$, the gradient in the y direction. The magnitude of vector is denoted by M (x, y), where:

$$M(x, y) = \text{mag}(\nabla f) = \sqrt{{g_x}^2 + {g_y}^2} \tag{3.12}$$

(Sochevanov, 1961; Safari et al., 2018). From this approach, background effects on geochemical anomalies are reduced (Gonzalez and Woods, 2008; Ziaii, 2008; Safari et al., 2018).

3.2.5 Fractal models

Fractal models are means and ways designed to explain complex behaviors or relationship existing between materials of the earth or universe (e.g., Turcotte, 1997). Unknowingly, human beings have been working with fractals in nature. Later in the 19th century, the Algerian-born mathematician Gaston Maurice Julia and latter on the Polish-born scientist Benoit Mandelbrot spent their lives studying the iteration of polynomials and rational functions on these complex relationships of materials and their shapes including the complex relationship between coastal lines, landscapes (e.g., Fig. 3.14, etc.) (www.fractal.org/Bewustzijns-Besturings-Model/Fractals-Useful-Beauty.htm; Agterberg, 2012). It is well established that fractal signals can be used to model natural objects, giving a room for mathematical definition of our environment with a little higher accuracy than before (e.g., Turcotte, 1997).

FIGURE 3.14 Several computer-generated fractal objects that show a remarkable resemblance to naturally occurring flora (Rouvray, 1996).

The concept of fractals was conceived for the first time by Cantor, Peano, Fatou, and others whereby the word "fractal" is an adjective for a Latin word "fructus" that corresponds to a Latin verb *frungere* which means "to break" or to create irregular fragments (Anderson et al., 1998).

The study of fractals is now the modern way of visualizing what was thought as impossible in mathematics, physics, geophysics, astronomy, and geology into simplified graphical images that are easily interpreted. In geology for example, particularly in exploration geochemistry, there are nearly a dozen of fractal models already in application. Some of these models include concentration–area, concentration–perimeter, concentration–number, concentration–volume, concentration–distance, number–size, spectrum–area, and local singularity analysis fractal models (Cheng et al., 2010; Zuo et al., 2012).

The fractal models are actually multifractal models—they are spatially intertwined (Mandelbrot, 1983; Stanley and Meakin, 1988) and their analysis is based on the self-similarity—an assumption that allows for prediction of strong local continuity of element concentration values that cannot be readily determined by variogram or correlogram analysis (Agterberg, 2012). The fractal geometry is scale-invariant, meaning that the properties of an object within a specific scale are somewhat similar to the properties of the same object within a different scale (Feder, 1988). Multifractal models are powerful tools for identifying geochemical anomalies and determining baselines for various geochemical surveys (Zuo et al., 2012). The main advantage of multifractal modeling is that in working with the data, there is no need to remove outliers (Khalajmasoumi et al., 2017).

3.2.5.1 Concentration area (C–A)

The concentration area model was first expressed by Cheng et al. (1994a) and is based on the distribution of geochemical elements concentration and how these distributions are related to the area enclosed by concentration contours.

The concentration area model is defined by the form:

$$A\,(\rho \leq \upsilon) \propto \rho^{-a1};\; A\,(\rho \geq \upsilon) \propto \rho^{-a2} \tag{3.13}$$

where A $(\rho \leq \upsilon)$ and A $(\rho \geq \upsilon)$ represent areas (A) with concentration values smaller or greater than contour value ρ; υ represents the threshold; $-a1$ and $-a2$ represent characteristic exponents which denote fractal dimension. Threshold values in this model indicate boundaries between different geochemical populations (Navidi et al., 2014; Zuo and Wang, 2016) (Fig. 3.15).

Threshold values are obtained by looking at log–log plots of concentration contours (that represent concentration values C) versus areas (that represent area of cells A), in which case certain concentration contours representing breakpoints in the plots are considered threshold values separating geochemical populations in the data (e.g., Zuo et al., 2009; Ajayebi et al., 2016; Khalajmasoumi et al., 2017). Distinct patterns, each corresponding to a set of similarly shaped contours, can be separated by different straight segments fitted to the values of the contours and enclosed areas on the log–log plots. The slopes of these straight lines can be taken as estimations of the exponents of the power–law relation (Eq. 3.13). The enclosed area can be calculated based on the contour map created by interpolation procedures, and based on superimposing a grid with cells on the study area and calculating the area by means of a box-counting method (Zuo and Wang, 2016). The threshold values of the data can then be taken from straight line segments fitted on the C–A log–log plot (Khalajmasoumi et al., 2017) as cross-points of the lines fitted to separate geochemical data into different populations,

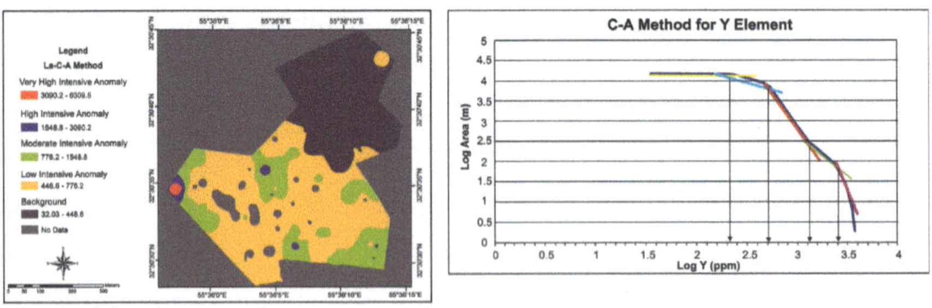

FIGURE 3.15 C-A log-log plots for Y along with its geochemical anomaly map (Khalajmasoumi et al., 2017).

which may represent different factors such as rock types, mineralization, alteration and surficial geochemical processes (e.g., Goncalves, 2001; Afzal et al., 2010; Khalajmasoumi et al., 2017).

3.2.5.1.1 Limitation of C−A model

The C−A model is not useful within a region with a complex geological setting and its results towards the edges of the maps are sometimes not very good, that is, the model can have edge effects (Daya and Afzal, 2015). Daya and Afzal (2015) also show that CA anomalous areas computed using C−A model are somewhat scattered and the boundaries of its anomalous areas relatively defuse.

3.2.5.1.2 Concentration number (C−N)

The C−N fractal model is a subset of C−A (Mao et al., 2004; Hassanpour and Afzal, 2013). It is also used to identify geochemical background, threshold and anomaly values (Mandelbrot, 1983; Zuo et al., 2009; Sadeghi et al., 2013; Hassanpour and Afzal, 2013; Afzal et al., 2014, 2016; 2017). The model operates based on an inverse relationship between elemental concentration and cumulative frequency of samples (Wang et al., 2007; Zuo et al., 2009; Hassanpour and Afzal, 2013). This model is presented by the following equation:

$$N(\geq \rho) \propto \rho^{-\beta} \tag{3.14}$$

where $N(\geq\rho)$ demonstrates the sample number with concentration values greater or equal to ρ. The ρ and β are concentration of ore element and fractal dimension, respectively. Unlike other models, pretreatment and evaluation of raw data before calculations are undertaken is not necessary with C−N (Hassanpour and Afzal, 2013). Eq. (3.14) can also be rewritten as:

$$\mathrm{Log}[N(\geq \rho)] = -\beta \mathrm{Log}(\rho) + \mathrm{Log}(C) \tag{3.15}$$

where C is the constant (Darabi and Hezarkha, 2018).

3.2.5.2 Spectrum area model (S−A)

Cheng (1999) proposed the S−A fractal model to represent the power-law frequency distribution of the power spectrum density for separating geochemical background from anomalies. In the following year, Prof. Cheng with colleagues published an article that showed the relationship between spectrum-energy-density (S) and the "area" of the set A ($\geq$S) with spectral−energy−density values above S on the power-density plane as:

$$A\,(\geq \mathrm{S}) \propto \mathrm{S}^{-2d/s} \tag{3.16}$$

where $2d$ is related to the so-called elliptical dimension with $d = 1$, corresponding to isotropic dimension, and $s > 0$, the exponent of the power law (Fyzollahhi et al., 2018).

In a log–log plot, straight lines can be drawn before an irregular fractal filter and can be built according to distinct patterns recognized by fitting these lines. In this process, every line has a different slope (Zuo and Wang, 2016). The geochemical anomalies of interest can be obtained by converting the filtered pattern back to the spatial domain (e.g., Cheng et al., 2010; Zuo and Wang, 2016)

3.2.5.2.1 Limitation of S–A model

The S-A anomaly maps are sometimes influenced by edge effects though the edge effects in an irregular shaped study area can be further investigated by taking more samples and getting more information at the edge of the area in question (Daya and Afzal, 2015).

3.2.5.3 Local singularity analysis model (L–SA)

Singularity is a fundamental property of complex natural processes such as earthquakes, mineralization, etc. (e.g., Zuo and Wang, 2016). It is a technique used to identify geochemical anomalies that result from singular processes (Zuo et al., 2009). This method can depict weak geochemical anomalies (Zuo et al., 2012; Yuan et al., 2015). The recent advances for identification of weak geochemical anomalies refer to singularity mapping technique proposed by Cheng (2007), and has been demonstrated as a powerful multifractal tool to recognize weak geochemical anomalies in complex geological settings or in overburden covered areas (e.g., Agterberg, 2012; Bai et al., 2010; Cheng, 2007; Zuo et al., 2015). Cheng (2007) show that:

$$\rho[\mathrm{Bx}(\varepsilon)] = \frac{\mu[\mathrm{Bx}(\varepsilon)]}{\varepsilon^E} = c(x).\varepsilon^{\alpha(x)-E} \tag{3.17}$$

where $\rho[\mathrm{Bx}(\varepsilon)]$ represents element concentration value determined on a neighborhood size measure Bx at point x, $\mu[\mathrm{Bx}(\varepsilon)]$ represents amount of metal, $c(x)$ is a coefficient representing smoothed element concentration value, and E is the Euclidean dimension of the sampling space.

In local Singularity analysis geochemical data collected at points are subjected to two treatments; (1) construction of contour maps by kriging or inverse distance weighting techniques or any other method generally used for this purpose and (2) the same data are subjected to local singularity mapping. The local singularity α is then used to enhance the contour map by multiplication of the contour value by the factor $\varepsilon^{\alpha-2}$ where $\varepsilon < 1$ represents a length measure and the factor $\varepsilon^{\alpha-2}$ is >2 in places where there has been local element enrichment or by a factor <2 where there has been local depletion. Local singularity mapping can be useful for the detection of geochemical anomalies characterized by local enrichment even if contour maps for representing average variability are not constructed (cf. Cheng and Agterberg, 2009; Zuo et al., 2009; Yuan et al., 2015). The technique can be applied on either raw or gridded data. For areas with irregularly spaced sampling, gridding the data allows for better visualization of the spatial characteristics of the data set and improves the accuracy of singularity mapping (Yuan et al., 2015; Parsa et al., 2017).

3.2.5.3.1 Limitation of local singularity analysis model

Its original algorithm used to estimate the local singularity exponent could not directly process a data set containing negative values and that the estimated singularity is influenced by background values (Zuo and Wang, 2016). However, this is currently overcome by modified algorithm that involves the introduction of a step in which the minimum value within the maximum window for each given location is previously subtracted—a straight processing technique that solves the existing problem and enables the singularity of similar anomalous patterns within different background values to be comparable (Zuo et al., 2015).

3.2.5.4 *Concentration perimenter (C−P), concentration volume (C−V), concentration distance (C−D), and number size (N−S)*

In this subsection, the C−P, C−V, C−D, and N−S fractal models are briefly discussed. Limitations on the applications of the models are not highlighted.

C−P: The concentration perimeter (C−P) model is also used to delineate geochemical anomalies from background values expressed as:

$$P(\geq \rho) \propto F\rho^{-\beta} \tag{3.18}$$

where ρ and P ($\geq\rho$) represent elemental concentration with concentration values greater than or equal to ρ respectively. F and β are a constant and the fractal dimension of the distribution of elemental concentrations, respectively (Cheng et al., 1994b; Cheng, 1995; Afzal et al., 2017).

C−V: On the basis of the same idea as the C−A model, the C−V fractal model was proposed by Afzal et al. (2010) to quantify the relationship between element concentration and the accumulative volume with concentration greater than or equal to the given value (Afzal et al., 2010; Delavar et al., 2012; Sadeghi et al., 2013; Soltani et al., 2014; Zuo and Wang, 2016).

C−D: A geochemical dispersion pattern often involves many subpatterns at many hierarchical levels, leading to the spatial distribution of element concentration being clustered at different scales (Li et al., 2003). Mandelbrot proposed the radial-density model as an approach to characterize the clustering of point events (Mandelbrot, 1983). Based on this model, Li et al. (2003) developed the concentration–distance (C–D) fractal model by replacing the density with element concentration. This model can directly process original element concentration data, and can avoid the error caused by any interpolation procedure. Determination of the optimum threshold for this model is nearly the same as for the C–A model.

The very important fractal model, that is, the number–size model (N–S), was firstly proposed by Mandelbrot (1983) to characterize the relationship between the size of objects and the number of objects with size greater than or equal to a given size. Based on this model, several variants have been developed and successfully applied in earth sciences (Agterberg, 1995; Carlson, 1991; Hassanpour and Afzal, 2013; Turcotte, 1997; Zuo et al., 2009). This model was probably the first one with geochemical relevance (Zuo and Wang, 2016).

3.2.6 Remote sensing geochemistry

3.2.6.1 Basic principles

Remote sensing is an emerging science useful in almost all fields (e.g., Gandhi and Sarkar, 2016); from geology/and mineral exploration (e.g., Hunt, 1980; Sabins, 1999; Ramanaidou et al., 2015) where images are used to map lithological units, faults and fractures in order to locate ore deposits through recognizing hydrothermally altered rocks following their spectral signatures (e.g., Sabins, 1999), land use and cover (e.g., Lin, 2013), military, archeology, agriculture (Buschmann and Nagel, 1993), surveying, geomorphology, deforestation, water quality management and dynamics, urban planning, air pollution (e.g., Maya et al., 2015), etc.

Principally, remote sensing is the collection of data from an object without being in physical contact with that object (Lillesand and Kiefer, 1994; Gandhi and Sarkar, 2016). It is usually undertaken using aircrafts (Sabins, 1999) spaceborne vehicles i.e. satellites such as Landsat Thematic Mapper (Landsat TM), and Satellite Pour l'Observation de la Terra (SPOT), airborne vehicles and nowadays using drones (e.g., Sabins, 1999; Wash, 2003). In order to collect the data, one needs to have a sensor. This produces image data that are observed, measured and stored in a spatial data base (Bakker et al., 2001). Remote sensing is undertaken using nearly the whole of the electromagnetic spectrum; that is, from low-frequency radio waves through the microwave, sub-millimeter, far infrared, near infrared, visible, ultraviolet, x-ray, and gamma-ray regions of the spectrum (e.g., Gibson, 2000; Elachi and van Zyl, 2006).

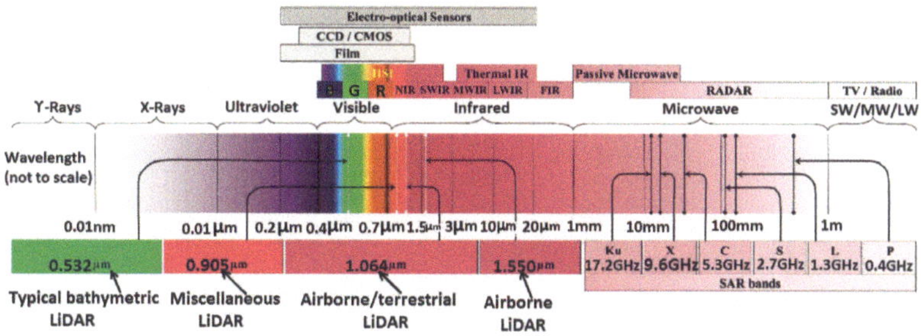

FIGURE 3.16 Electromagnetic spectrum with important remote sensing bands. Note the visible light spectrum is from 0.4 μm to 0.7 μm (i.e., Violet: 0.4–0.446 μm, Blue: 0.446–0.500 μm, Green: 0.500–0.578 μm, Yellow: 0.578–0.592 μm, Orange: 0.592–0.620 and Red: 0.620–0.7 μm; www.nrcan.gc.ca/node/14623#answer). Other spectrums (x-rays, ultraviolet, Infrared and its sub-sets (NIR, MIR, FIR), microwave and radio waves) are shown (Toth and Jóźków, 2016).

Several factors characterize the study of remote sensing: (1) Electromagnetic spectrum, that is a window at which these remotely sensed data are measured (Fig. 3.16); (2) atmospheric effects,—that interacts/interferes with the waves carrying the image signal; (3) Properties of Target objects, that is whether or not the target has a color that is bright or not, has a rough or smooth surface, has a white or black color, etc.; (4) configuration of the satellite or airborne platforms, (5) energy source, and (6) Spectral reflectance curves (Fig. 3.17)—all these factors affect the remote sensing data collection, quality and consequently interpretation of the same (e.g., Sabins, 1999).

Every object emits energy in the form of electromagnetic waves. These waves wonder in the atmosphere or/and on the earth's surface interacting with atmospheric particles such as gases, dusts and water vapors. They also interact with the earth's surface in multiple ways before they are reflected in the atmosphere (e.g., Hunt, 1980). Consequently, waves from an object can undergo both reflection and refraction leading to modification of their original energy content (e.g., Sabins, 1999). The object here can be the earth's surface itself, the water bodies, the atmosphere and its gasses, trees or vegetation, etc. (Sabins, 1983, 1997, 1999).

In simple terms, when a satellite records some energy in the form of electromagnetic waves, the principle is that:

- Electromagnetic energy is emitted from the source, such as the sun to various objects that reflect / and refract the energy in various directions (Fig. 3.17),
- Electromagnetic energy is then detected and recorded by a sensor before data is displayed on computers for visual and numerical interpretations.

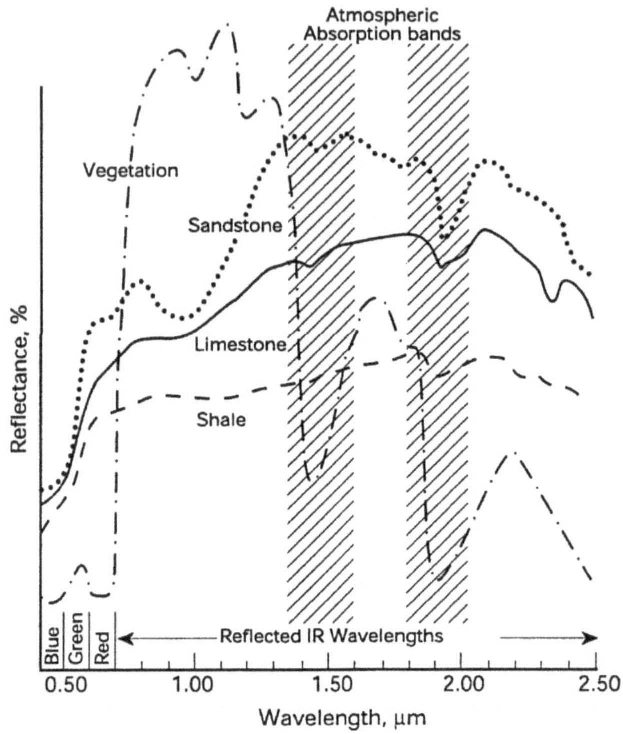

FIGURE 3.17 Spectral bands recorded by remote sensing systems. Spectral reflectance curves are for vegetation and sedimentary rocks. Source: *From Sabins, F.F., 1999. Remote sensing for mineral exploration. Ore Geol. Rev. 14, 157-183.*

When the returned energy from the object is directly measured by the satellite which supplies the energy source, it is referred to as active arrangement (e.g., Mason, 2005) and when the light has to pass through another body, get reflected before it is detected by the sensor, the arrangement is passive (Mason, 2005; Fig. 3.18).

In the active arrangement, the principle is that the satellite or radar provides its own energy that goes directly to the target object as in Fig. 3.18. For the passive arrangement, the relationship between object, source of energy (e.g., sun), object, sensor, and light is in accordance to Eq. (3.19):

$$M_r(\lambda) = \rho(\lambda)E(\lambda) \tag{3.19}$$

where $M_r(\lambda)$ is the reflected solar radiation of spectral length λ by the land surface, or an image of spectral band λ, $E(\lambda)$ is irradiance, the incident solar radiation energy upon the land surface, while $\rho(\lambda)$ is the reflectance of land surface at wavelength λ (Mason, 2005).

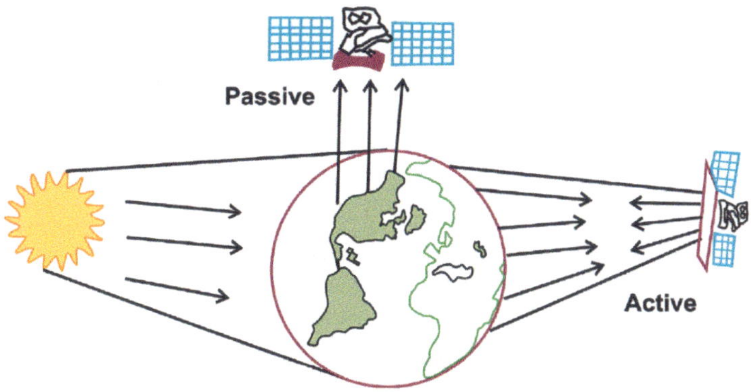

FIGURE 3.18 Schematic diagram showing the position of a data collecting device with respect to Sun and Earth to designate it as a passive or active sensor (Haldar, 2018—Chapter 3).

Every matter (material) has its own spectral behavior when illuminated by light. Some materials absorb light and some reflect light at different angles and intensities. Some materials would reflect part of the light rays and refract some within the visible spectrum or outside the visible spectrum (e.g., Buschmann and Nagel, 1993). Chlorophyll for example would absorb the blue light and part of the red light reflecting the green light almost wholly making the plant look green. In other words, the incidence blue and red wavelengths are absorbed by chlorophyll in the process of photosynthesis whereas the incident green wavelengths are partially reflected by chlorophyll (Buschmann and Nagel, 1993; Gitelson et al., 1998).

3.2.6.2 Application in mineral exploration: alteration minerals

Applications of remote sensing are diverse. In mineral exploration, they are more useful in identifying mineral alteration zones, which are the manifestation of geochemical reactions between rock-forming minerals and hydrothermal fluids. The more the alteration minerals, the more the geochemical reactions they reflect (e.g., Sabins, 1987, 1997), and therefore, the more potential for mineral deposit the area is (e.g., Siegal and Gillespie, 1980). Each mineral deposit has its characteristic alteration minerals that characterize that particular deposit. For stance, ferruginization (i.e. Fe-alteration) is the function of alteration of iron-rich rocks close to the surface (e.g., MüCke, 1994), sericitization is associated with hydrothermal alteration of plagioglase within the phyllic zone and kaolinization also indicates plagioclase or feldspar alterations.

Different alteration minerals display different signatures that result from distinctive effects they cause to the incoming electromagnetic radiations. Further, as there are different types of satellites within

ASTER satellites family, their bands, even when assigned same number, are different and so do sub-types within the Landsat satellites (Rekhibi et al., 2015). The applications of various bands even those that are assigned same numbers are also different. For instance, Band 1 from ASTER (VNIR-Band 1) has a wavelength 0.520 to 0.600 with resolution of 15 m. Band 1 from Landsat 4–5 Thematic Mapper (TM) has wavelength 0.45 to 0.52 (e.g., Barsi et al., 2014) and is useful in Bathymetric mapping, distinguishing soil from vegetation, and deciduous from coniferous vegetation (e.g., Barsi et al., 2014).

The RGB (i.e., Red–Green–Blue) color bands from ASTER satellites can be used as follows: (RGB) = bands 3, 2, 1 (expressed as 321) is used to identify alteration mineral areas (e.g., Brandmeier et al., 2013). RGB = 741 (742) is used for identification of alteration mineral areas (e.g., Sabins, 1987; Brandmeier et al., 2013), RGB = 5/7 3/1 3/4 is a Band ratio used in mapping minerals/alteration minerals (e.g., Gabr et al., 2015) and similarly, RGB = 3/1 2/3 3/4 is a band ratio used in mapping minerals/alteration minerals (e.g., Sadiya et al., 2014; Ali et al., 2017). Further, in ASTER satellite data processing, 321 images, 468 images, 101214 image, DEM image and band ratio image are also used to find areas of alteration. For instance, kaolinite, allunite, halloysite, sericite (muscovite), illite and smectite can be easily identified within VNIR as white color using 321 band. When the color is yellowish white it could imply kaolinite, allunite, halloysite, sericite (muscovite), illite and smectite (e.g., Pour and Hashim, 2012; Sakar et al., 2019).

Both Landsat 08 and ASTER (VNIR and SWIR) can be acquired and processed to generate different band ratios to highlight different types of mineral alteration zones and thereby throwing light on ore formation system and structural controls of ore localization. Band ratios from VNIR and SWIR ASTER bands for different minerals, through the Iron Oxides Index; Alunite Index (ALI); Kaolinite Index (KLI); and Muscovite Index have been used to demarcate the alteration zones (Pour and Hashim, 2012; Ali et al., 2017):

$$\text{Iron oxides Index} = \frac{Band\ 2}{Band\ 1} \tag{3.20}$$

$$\text{ALI} = \frac{(Band\ 7)}{(Band\ 5)} * \left(\frac{Band\ 7}{Band\ 8}\right) \tag{3.21}$$

$$\text{KLI} = \frac{(Band\ 4)}{(Band\ 5)} * \left(\frac{Band\ 8}{Band\ 6}\right) \tag{3.22}$$

$$\text{Muscovite Index} = \frac{Band\ 7}{Band\ 6} \tag{3.23}$$

3.2.6.2.1 Limitation of remote sensing

By comparison, ASTER, apart from having more bands than those of LANDSAT, it also has higher resolution than LANDSAT which means it is able to show more alteration minerals than LANDSAT. Further, ASTER sensor consists of three separate subsystems and has a sun-synchronous polar orbit meaning it crosses over any given latitude at the same time each day and it revisits the same area every 16 days.

Apart from the power of remote sensing in day-to-day scientific operations, particularly in earth sciences, yet, technologies for effective use of the data and for extracting useful information from the data of Remote sensing are still very limited since no single sensor combines the optimal spectral, spatial and temporal resolution (Al-Wassai and Kalyankar, 2013).

3.2.7 The use of handheld/portable XRF (pXRF)

Unlike the laboratory XRF that uses alternating currents to emit high energy photons that strikes sample atoms in the given sample (as incident rays) so that electrons from the shells are dislodged, the source of power for pXRF is direct current batteries. The invention of the pXRF is a result of miniaturization of the machine to suit exploration activities outside a conventional laboratory. With small size and lightweight, pXRF can be moved anywhere. Analysis by portable XRF gives a very good picture of elemental composition of material in the field which is comparable with conventional laboratory results. For some elements, portable XRF results are exactly the same to results from laboratory XRF or any other technique such as AAS and even ICP (Seneshen, 2015, Fig. 3.19). Exceptions exist when lower detection limits are considered.

The pXRF such as InnobXsytems and Thermo's Niton XL EDXRF measure element concentrations from ^{12}Mg to ^{92}U. Typically measures 1 × 1 cm area to a depth of about 2 mm in soil. It is an ideal instrument for Pb and EPA approved for environmental surveys. The pXRF has made life easy in mineral exploration as this machine is portable and can be used in the field much easier and quicker (Bosco, 2013; Moon and King, 2015).

The fact that the pXRF is quicker, and the fact that its accuracy is comparable with the laboratory XRF machine, exploration programs have been made comparably quicker. For instance, Le Vaillant et al. (2018) use pXRF to analyze for Ni, Cr, Ti and other compatible elements in the field and be able to immediately plot the log Ni/Ti versus log Ni/Cr and discriminate fertile from barren komatiites in the Yilgan craton, Australia (Fig. 3.20).

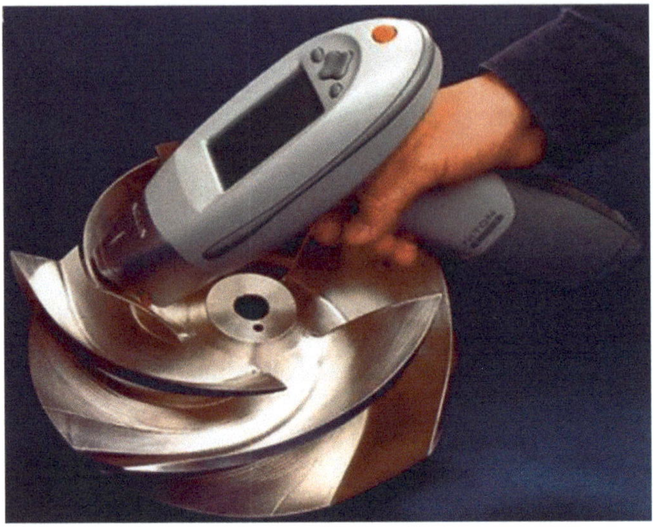

FIGURE 3.19 The new portable XRF directed on an object. Note the "gun" design appearance (Bosco, 2013).

3.2.8 The use of mineral prospectivity modeling

The continuous improvement of geological exploration techniques and methods has led to a significant increase in the amount and type of geoscientific data that is used in mineral exploration. However, this increase in data availability has also generated a number of challenges in terms of how to extract and combine only those data that can effectively be used to predict the location of hitherto undiscovered mineralization (Harris, 1984; Agterberg et al., 1990; Carranza, 2009). Mineral prospectivity modeling is an effective method of target identification during mineral exploration and involves the integration and combination of multiple geoscientific datasets (Agterberg et al., 1990; Bonham-Carter, 1994). Mineral prospectivity modeling using geoscientific data such as geological, geochemical, geophysical and remote sensing data, among others, has been widely used for mineral exploration and targeting for the last 30 years (Carranza, 2009; Carranza and Sadeghi, 2010; Ford and Blenkinsop, 2008; Lindsay et al., 2014; Pan and Harris, 2000; Porwal et al., 2010).

In recent years, 3D mineral prospectivity modeling approaches and multifractal method based mineral prospectivity modeling have been used for exploration for concealed and deep-seated mineralization (Chen et al., 2007; Fallara et al., 2006; Li et al., 2015, 2019; Sprague et al., 2006; Wang et al., 2011; Yuan et al., 2014a,b; Payne et al., 2015). In Chapter 4, several case studies will be introduced to show the

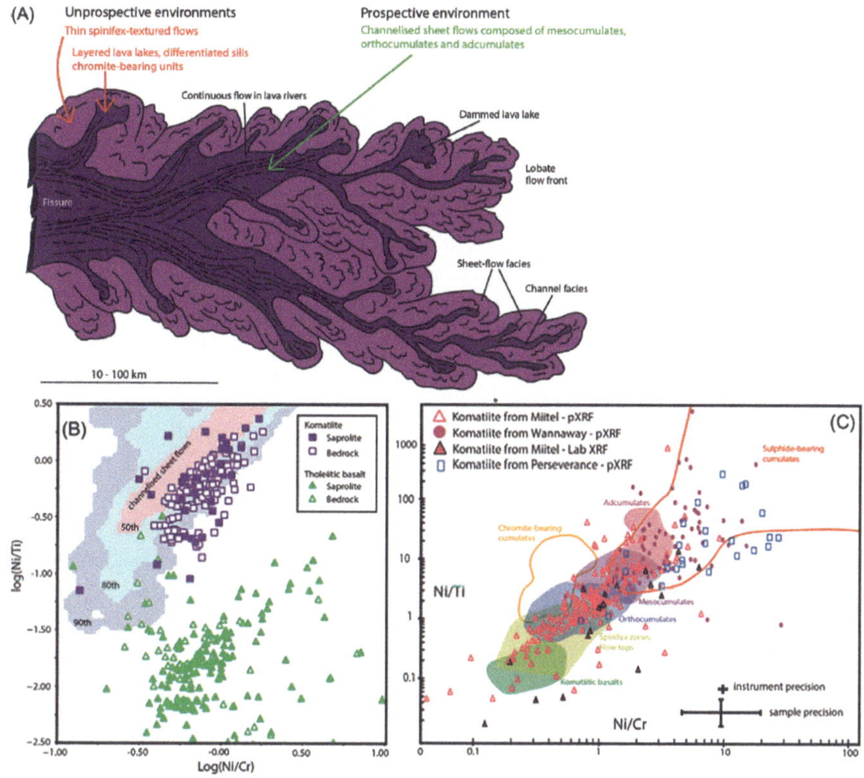

FIGURE 3.20 (A) Schematic illustration of a regional komatiite flow field, modified from Hill et al. (1995), (B) comparison of Ni/Ti and Ni/Cr ratios between fresh bedrock and "top of fresh rock" saprolite in the Agnew area in Western Australia, data compiled by Barnes et al. (2014), and (C) Plot showing the potential use of pXRF in evaluating the prospectivity of a komatiite unit using Ni/TI and Ni/Cr ratios (After Le Vaillant et al., 2018).

approaches and workflow of 3D mineral prospectivity modeling and multifractal method based mineral prospectivity modeling.

3.2.8.1 3D—prospectivity modeling

The targeting of deep and concealed ore deposits during mineral exploration is extremely challenging, primarily as a result of the inherent increase in complexity when exploring in deeper areas with less dense geological information. One approach that can provide new insights into exploration targeting at depth is the analysis and modeling of 3D geological objects using an integrated geophysical and geological approach (Lindsay et al., 2014; Li et al., 2015). These 3D mineral prospectivity modeling approaches are based on 3D geological models and have been applied to the exploration targeting of deep-seated and concealed mineralization elsewhere both on regional (Li et al., 2015, 2019; Nielsen et al., 2015; Perrouty et al., 2014; Payne et al., 2015) and deposit (Wang and Huang, 2012; Yuan et al., 2014a; Xiao et al., 2015) scales.

The workflow and methods used in 3D mineral prospectivity modeling can be summarized to four steps (Sprague et al., 2006; Mao et al., 2011; Xiao et al., 2012; Yuan et al., 2014b, 2018): (1) construction of 3D geological database; (2) 3D geological modeling, (3) 3D predictive map generation and (4) 3D data fusion and mineral prospectivity mapping (Fig. 3.21). The first step collects and digitizes the geological data, such as geological maps, cross-sections, boreholes, and geophysical data, and then designs and constructs the 3D geological database. Based on the 3D geological database, the second step uses available geological and geophysical data to construct 3D geological models, such as sedimentary unit, intrusion and fault models. In this step, initial joint geophysical inversion and forwarding modeling and 3D geological modeling approaches are used to combine available geological and geophysical data within a 3D geological model of the study area. The 3D geological models are then used to quantify the exploration criteria identified in the exploration model, leading to the generation of 3D predictive maps for each criterion using 3D spatial analysis, 3D computational simulation and 3D geophysical methods. The final step in this process involves combining all of the 3D predictive maps created in step three to create a 3D prospectivity map that enables the delineation of exploration targets and estimation of the resources for mineralization in the study area. Hereinafter, several case studies will be used to show the methodology, workflow and results of three-dimensional mineral prospectivity modeling within ore-field and deposit scale.

3.2.9 Whole-rock analysis for incompatible elements Sr and Y

Whole-rock analysis can be a powerful tool to discriminate fertile from barren environments. For instance, porphyry Cu deposits tend to have higher Sr/Y ratios than barren environments. This is attributed to the fact that Cu—porphyry deposits are a result of repeated supply of water as well as repeated cycles of replenishment by primitive magma and partial crystallization—such processes tend to concentrate Cu (e.g., Rohrlach and Loucks, 2005). At the same time plagioclase is suppressed from being crystallized and instead hornblende crystals are crystallized (e.g., Cohen and Bowell, 2014). Plagioclase and hornblende crystals are richer in Sr and Y respectively resulting in higher Sr/Y ratios—some conditions that are optimal for Cu mineralization (Richards, 2011; Cohen and Bowell, 2014).

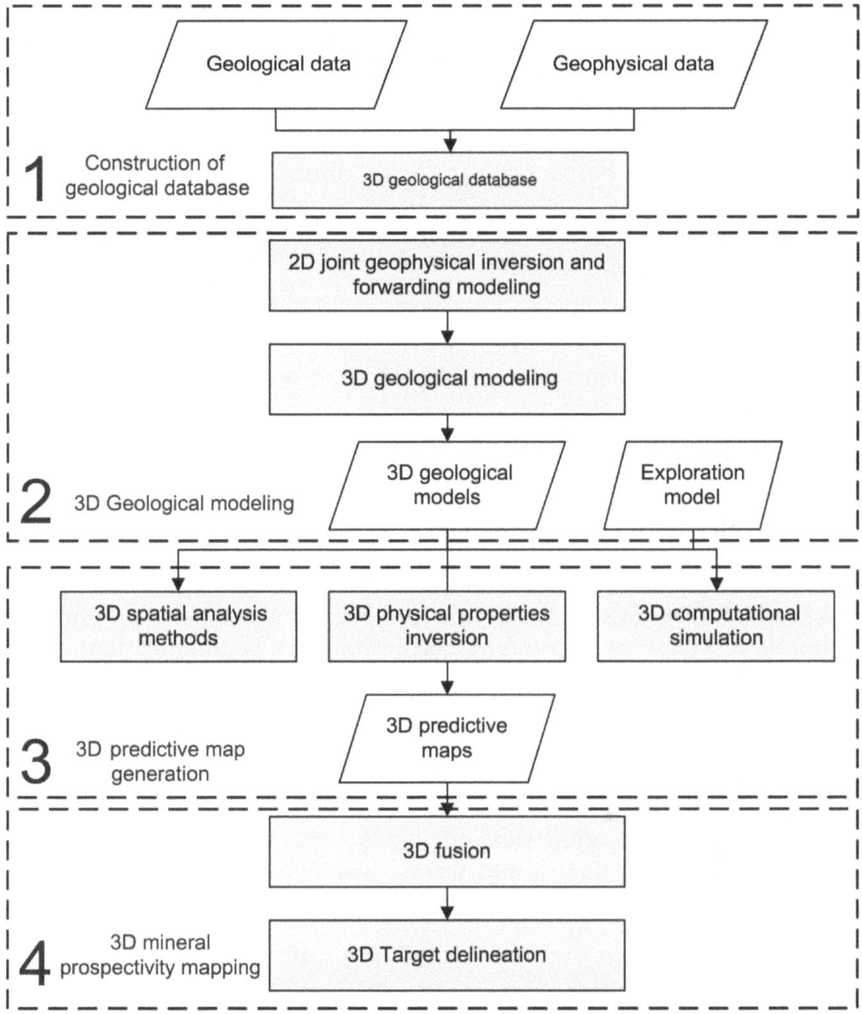

FIGURE 3.21 Workflow of three-dimensional prospectivity modeling. Source: *Modified after Yuan, F., Li X., Zhang M., Jia C., Hu X., 2018. Research progress of 3D prospectivity modeling. Gansu Geo-eng. 27, 32-36 (In Chinese with English abstract); Li, X., Yuan, F., Zhang, M., Jowitt, S.M., Ord, A., Zhou, T., et al., 2019. 3D computational simulation-based mineral prospectivity modeling for exploration for concealed Fe–Cu skarn-type mineralization within the Yueshan orefield, Anqing district, Anhui Province, China. Ore Geol. Rev. 105, 1-17.*

On the other hand, abundance, grain size, and chemical composition of rutile correspond to the intensity of alteration and mineralization (Scott, 2005). For instance, in Canada, V, W, and Sb within rutile have been used to distinguish intensity of mineralization at Hemlo gold deposit (Harris, 1984). It can be particularly useful where rutile is concentrated in regolith derived from a mineralized protolith in which the endogenic composition of the precursor mineralization is retained in the resistant rutile. Gold occurs as disseminated components along with abundant sulfides as 5–10 mm grains within pyrite, and rutile has a characteristic V + Fe + W + Sb geochemical signature in the gold-bearing zone which has a signature extending 200 m into the footwall. This feature provides a potentially larger target for detecting proximity to gold mineralization than the dispersion of gold in the protolith or regolith (O'Reilly et al., 2004).

3.2.10 Determination of number of geochemical samples

In order to save time and resources, an optimum number of samples to be taken/collected in the field has to be established. It has conventionally been taken that 30 or more soil or rock samples are enough to provide room for meaningful statistical computations. In reality, one may collect any number of samples but chances to generate ore target(s) are unknown (e.g., Kleeman, 1967). What inspires exploration geologists/geochemists is the assurance to get an anomaly (from which an economic ore body can be intersected). There are techniques that geologists use to establish the optimum samples to be collected in the field. Kleeman (1967) and Govett (1983) agree that 10^6 to 10^7 particles of a sample is the required quantity at the stage of analysis. Considering specific gravity of the sample to be 2.75 g/cm^3 in an aliquot of 1 g, then the sample must be ground to 170 mesh (A.S.T.M). According to Govett (2000), it is recommended that during the crushing and grinding, the total amount of samples should not be less than 160 g at 35 mesh and not less than 8 g at 80 mesh, in which case

$$g = \frac{2.75\pi d^3}{6}\left[\frac{(100-P)\times 10^4}{PR_{\mathrm{a}}^2}\right] \tag{3.24}$$

where d stands for the grain diameter, R is relative standard deviation, P is percentage (Govett, 1983). This expression is represented graphically in Fig. 3.22.

According to Govett (1983), $N = (p, \mathrm{Pr} > r)$; where N = Number of samples required to yield at least "r" samples with element contents (or element ratios or other parameters) $>/<$ predetermined value with a probability of occurrence of "p."

Govett shows that we need 200 for samples for at least three anomalous samples, and if we need at least five anomalous samples, then 300 samples have to be collected. We also need 100 samples to obtain at least one anomalous sample (Table 3.3).

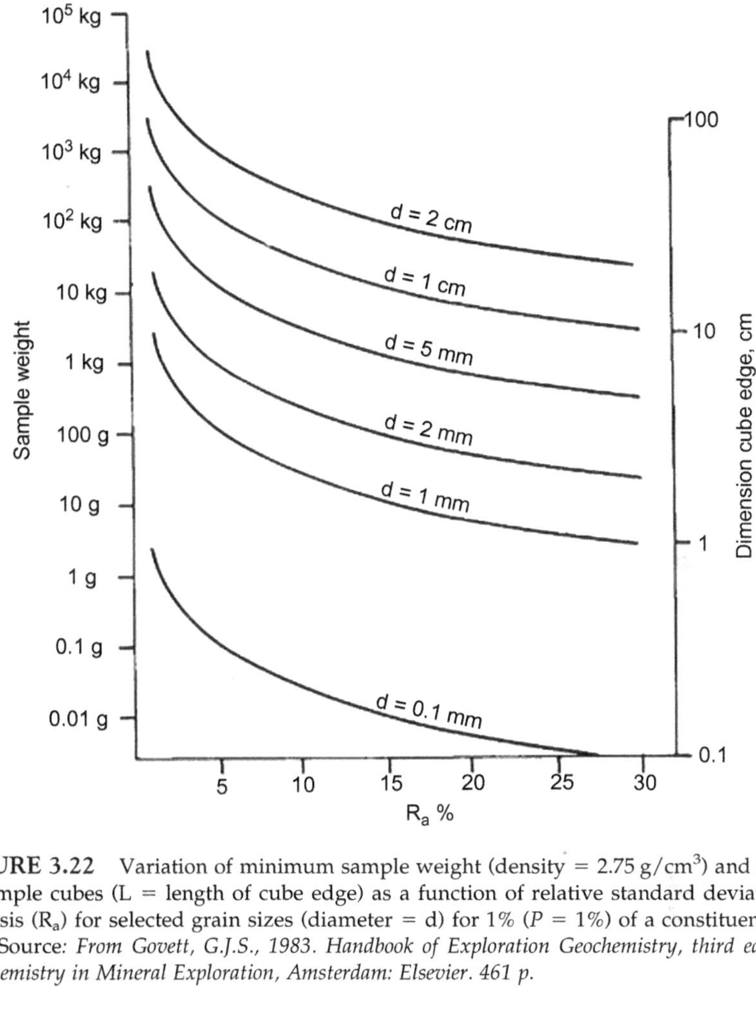

FIGURE 3.22 Variation of minimum sample weight (density = 2.75 g/cm^3) and in size of sample cubes (L = length of cube edge) as a function of relative standard deviation of analysis (R_a) for selected grain sizes (diameter = d) for 1% (P = 1%) of a constituent mineral. Source: *From Govett, G.J.S., 1983. Handbook of Exploration Geochemistry, third ed. Rock Geochemistry in Mineral Exploration, Amsterdam: Elsevier. 461 p.*

TABLE 3.3 Relationship between number of samples to be collected N, versus probability of occurrence, p.

Number of samples, *r*, > (or <)	Probability of occurrence, *p*									
	0.01	0.03	0.05	0.08	0.1	0.14	0.2	0.3	0.4	0.5
1	299	99	59	36	29	20	14	9	6	5
3	628	208	124	77	61	43	30	19	14	11
5	913	303	181	112	89	63	44	28	21	16
7	1182	392	234	146	116	82	57	37	27	21
10	1568	521	311	193	154	109	76	49	36	28

From Govett, G.J.S., 1983. Handbook of Exploration Geochemistry, third ed. Rock Geochemistry in Mineral Exploration, Amsterdam: Elsevier. 461 p.

3.2.10.1 *How to obtain probability of occurrence,* p

There could be various ways to obtain probability of occurrence of an anomalous value within a data set. Here we introduce a simplified approach. Suppose after plotting cumulative frequency against class intervals of data obtained from orientation survey, two populations are obtained, i.e. anomalous and nonanomalous values. Suppose the anomalous values are 30 out of 1000 total data (*N*). Then probability of occurrence of an anomalous value in this data would be 30/1000 = 0.03.

It follows therefore that, as per Table 3.3, if we need to obtain at least 7 anomalous samples in the successive missions, we need to collect at least 392 in total. If we need to obtain 10 anomalous samples from the data set, then we need to collect at least 521.

3.3 Concluding remarks

This Chapter highlights few conventional and nonconventional mineral exploration techniques. Nonconventional exploration techniques include Mineral prospectivity mapping, Lithogeochemistry, Mineral zoning, Staged factor analysis, Concentration gradient modeling, fractal models, the use of Mineral Prospectivity modeling and Remote sensing. The use of portable XRF (Pxrf), whole rock analysis for incompatible elements Sr and Y and determination of number of geochemical samples have been added in the list of nonconventional techniques used in mineral exploration. The list is not exhaustive. What is important here is to understand the principle (s) behind each of the methods and the limitation (s) associated with the technique before applying it. In case a technique is applied in an area where it is not recommended, chances to miss an orebody are high and where the unrecommended technique becomes successful it does not mean that the recommendation is wrong but that some other parameters on its application were not fully known. This means, as we continue to apply conventional or nonconventional techniques, our minds have to be open always. But, it is an undisputable fact that nonconventional exploration techniques are the ones that will make our world continue to have sufficient supply of the mineral resources that are continuously decreasing versus geometrically growing world population.

References

Afzal, P., Khakzad, A., Moarefvand, P., Rashidnejad, O.N., Esfandiari, B., Fadakar, A.Y., 2010. Geochemical anomaly separation by multifractal modeling in Kahang (GorGor) porphyry system, Central Iran. J. Geochem. Explor. 104, 34–46.

Afzal, P., Zarifi, A., Sadeghi, B., 2014. Delineation of geochemical anomalies using factor analysis and concentration-number (C-N) fractal modeling based on stream sediments data in Esfordi 1:100000 sheet, Central Iran. Iran. J. Earth Sci. 5, 100–110.

Afzal, P., Mirzaei, M., Yousefi, M., Adib, A., Khalajmasoumi, M., Zarifi, A.Z., et al., 2016. Delineation of geochemical anomalies based on stream sediment data utilizing fractal modeling and staged factor analysis. J. Afr. Earth Sci. 119, 139–149.

Afzal, P., Yasrebi, A.B., Saein, L.D., Panahi, S., 2017. Prospecting of Ni mineralization based on geochemical exploration in Iran. J. Geochem. Explor. 181, 294–304.

Agterberg, F.P., 1995. Multifractalmodeling of the sizes and grades of giant and supergiant deposits. Int. Geol. Rev. 37, 1–8.

Agterberg, F.P., 2012. Multifractals and geostatistics. J. Geochem. Explor. 122, 113–122.

Agterberg, F., Bonham-Carter, G., Wright, D., 1990. Statistical pattern integration for mineral exploration. In: Gaal, G., Merriam, D. (Eds.), 1990. Computer applications in resource estimation. Prediction and assessment for metals and petroleum. Computer and Geology, vol. 7. Pergamon Press, pp. 1–21.

Aitchison, J., 1986. The Statistical Analysis of Compositional Data. Chapman and Hall, London, 436 p.

Ajayebi, K., Jafari, H., Behbahani, B., 2016. The study of area-concentration fractal method in litho-geochemical data in Tanurjeh Area, Khorasan Province. Open J. Geol. 5, 451–457.

Albanese, S., De Vivo, B., Lima, A., Cicchella, D., 2007. Geochemical background and baseline values of toxic elements in stream sediments of Campania region (Italy). J. Geochem. Explor. 93, 21–34.

Ali, G.M.R., Khidir, S.O.H.E., Abdelgali, M.Y., 2017. Application of ASTER data in the identification of alteration zones-related to mineralization around Wadi El Milk, North Kordofan, Sudan. Al Neelain J. Geosci. 1 (1), 36–44.

Al-Wassai, F.A., Kalyankar, N.V., 2013. Major limitations of satellite images. J. Glob. Res. Comp. Sci. 4 (5), 51–59.

Ames, D.E., Franklin, J.M., Hannington, M.D., 1993. Mineralogy and geochemistry of active and inactive chimneys and massive sulphide, middle valley, northern Juan De Fuca Ridge: An evolving hydrothermal system. Can. Mineral. 31, 997–1024.

Anand, R.R., Aspandiar, M.F., Noble, R.R.P., 2016. A review of metal transfer mechanisms through transported cover with emphasis on the vadose zone within the Australian regolith. Ore Geol. Rev. 73, 394–416.

Anderson, A.N., McBramey, A.B., Crawford, J.W., 1998. Application of fractals to soil studies. Adv. Agron. 63, 1–76.

Atchley, W.R., 2007. Introduction to Principal Components and Factor Analysis (ftp://statgen.ncsu.edu/pub/thorne/molevoclass/AtchleyOct19.pdf).

Bagas, L., Xiao, K., Jessell, M., Li, N., 2017. Quantitative assessment of China's mineral resources Part 1. Ore Geol. Rev. 91, 1081–1083. Available from: https://doi.org/10.1016/j.oregeorev.2017.11.010.

Bai, J., Porwal, A., Hart, C., Ford, A., Yu, L., 2010. Mapping geochemical singularity using multifractal analysis: application to anomaly definition on stream sediments data from Funin Sheet, Yunnan, China. J. Geochem. Explor. 104, 1–11.

Bakker, W.H., Janssen, L.L.F., Reeves, C.V.B., Gorte, G.H., Pohl, C., Weir, M.J.C., et al., 2001. Principles of remote sensing. an introductory textbook. ITC Educational Textbook Series. Enschede, The Netherlands.

Barsi, J., Lee, K., Kvaran, G., Markham, B., Pedelty, J., 2014. The spectral response of the Landsat-8 operational land imager. Rem. Sens. 6 (10), 10232–10251.

Barnes, S.J., Fisher, L.A., Anand, R., Uemoto, T., 2014. Mapping bedrock lithologies throughin situregolith using retained element ratios: a case study from the Agnew-Lawlers area, Western Australia. Australian Journal of Earth Sciences 61, 269–285.

Basilevsky, A., 1994. Statistical Factor Analysis and Related Methods. Theory and applications. John Wiley and Sons, New York.
Beach, T., Ulmer, A., Cook, D., Brennan, M.L., Luzzadder-Beach, S., Doyle, C., et al., 2018. Geoarchaeology and tropical forest soil catenas of northwestern Belize. Quatern. Int. 463, 198–217. Available from: https://doi.org/10.1016/j.quaint.2017.02.031.
Billa, M., Cassard, D., Guillou-Frottier, L., Lips, A.L.W., Tourlière, B., 2002. Assessment of GIS Andes: predictive mapping of Neogene gold-bearing magmatic-hydrothermal systems in the Central Andes. 5th International Symposium on Andean Geodynamics, September 16–18, 2002, Toulouse, France. Extended Abstracts Volume. IRD Publications: Paris, 89–92.
Bonham-Carter, G., 1994. Geographic Information Systems for Geoscientists: Modelling with GIS. Elsevier, Oxford, 398 p.
Borg, G., Krogh, T., 1999. Isotopic age data of single zircons from the Archaæan Sukumaland Greenstone Belt, Tanzania. J. Afr. Earth Sci. 29 (2), 301–312. Available from: https://doi.org/10.1016/s0899-5362(99)00099-8.
Bosco, G.L., 2013. Development and application of portable, hand-held X-ray fluorescence spectrometers. TrAC Trends Analyt. Chem. 45, 121–134. Available from: https://doi.org/10.1016/j.trac.2013.01.006.
Buschmann, C., Nagel, E., 1993. In vivo spectroscopy and internal optics of leaves as basis for remote sensing of vegetation. Int. J. Remote. Sens. 14 (4), 711–722.
Brand, N.W., 1999. Element ratios in nickel sulphide exploration: vectoring towards ore Environments. J. Geochem. Explor. 67, 145–165.
Brandmeier, M., Erasmi, S., Hansen, C., Höweling, A., Nitzsche, K., Ohlendorf, T., et al., 2013. Mapping patterns of mineral alteration in volcanic terrains using ASTER data and field spectrometry in Southern Peru. J. South Am. Earth Sci. 48, 296–314. Available from: https://doi.org/10.1016/j.jsames.2013.09.011.
Cameron, E.M., Leybourne, M.I., Kelley, D.L., 2002. Exploring for deeply covered mineral deposits: formation of geochemical anomalies in northern Chile by earthquake induced surface flooding of mineralized ground waters. Geology 30, 1007–1010.
Carranza, E.J.M., 2009. Geochemical anomaly and mineral prospectivity mapping in GIS. In: Hale, M. (Ed.), Handbook of Exploration and Environmental Geochemistry, Vol. 11. Elsevier Science Ltd, Oxford, 368 p.
Carranza, E.J.M., Sadeghi, M., 2010. Predictive mapping of prospectivity and quantitative estimation of undiscovered VMS deposits in Skellefte district (Sweden). Ore Geol. Rev. 38, 219–241.
Carranza, E.J.M., Laborte, A.G., 2015. Data-driven predictive mapping of gold prospectivity, Baguio district, Ore Geol. Rev., 71. Application of Random Forests algorithm, Philippines, pp. 777–787.
Carlson, C.A., 1991. Spatial distribution of ore deposits. Geology 18, 111–114.
Cassard, D., Billa, M., Bouchot, V., Salleb, A., Stein, G., Tourlière, B., 2001. Predictive mapping with GIS Andes datasets. In: European Union of Geosciences, EUG XI, April 8–12, 2001, Strasbourg, France, J. Conf. Abs. 6(1), 265.
Cassard, D., Billa, M., Lambert, A., Picot, J.-C., Husson, Y., Lasserre, J.-L., et al., 2008. Gold predictivity mapping in French Guiana using an expert-guided data-driven approach based on a regional-scale GIS. Ore Geol. Rev. 34, 471–500.
Chen, J., Lu, P., Wu, W., Zhao, J., Hu, Q., 2007. A 3D method for predicting blind ore bodies, based on a 3D visualization model and its application. Earth Sci. Front. 14, 54–62.
Chen, S., Grunsky, E.C., Hattori, K., Liu, Y., 2015. Principal component analysis of geochemical data from the REE-rich Maw Zone, Athabasca Basin, Canada. Nat. Res. Can. Geological Survey of Canada—Open File 7689.
Cheng, Q., 1995. The perimeter–area fractal model and its application to geology. Math. Geol. 27, 69–82.

Cheng, Q., 1999. Multifractal interpolation. In: Lippard, S.J., Naess, A., Sinding-Larsen, R. T. (Eds.), Proc. Fifth Annual Conf. International Association of Mathematical Geosciences. Tapir Akademisk Forlag, Norway, pp. 245–250.

Cheng, Q., 2007. Mapping singularities with stream sediment geochemical data for prediction of undiscovered mineral deposits in Gejiu, Yunnan Province, China. Ore Geol. Rev. 32, 314–324.

Cheng, Q., Agterberg, F.P., 2009. Singularity analysis of ore-mineral and toxic trace elements in stream sediments. Comput. Geosci. 35, 234–244.

Cheng, Q., Bonham Carter, G.F., Agterberg, F.P., Wright, D.F., 1994a. Fractal modeling in the geosciences and implementation with GIS. Proceedings of the 6th Canadian Conference on GIS 1, 565–577.

Cheng, Q., Agterberg, F.P., Ballantyne, S.B., 1994b. The separation of geochemical anomalies from background by fractal methods. J. Geochem. Explor. 51, 109–130.

Cheng, Q., Agterberg, F.P., Bonham-Carter, G.F., 1996. A spatial analysis method for geochemical anomaly separation. J. Geochem. Explor. 56, 183–195.

Cheng, Q., Xia, Q., Li, W., Zhang, S., Chen, Z., Zuo, R., et al., 2010. Density/area power-law models for separating multi-scale anomalies of ore and toxic elements in stream sediments in Gejiumineral district, Yunnan Province, China. Biogeosciences 7, 3019–3025.

Child, D., 2006. The Essentials of Factor Analysis, third ed. Continuum International Publishing Group, New York, NY.

Clarke, F.W., 1924. The Data of Geochemistry, fifth ed. USGS Bulletin, pp. 770–841.

Cohen, D.R., Bowell, R.J., 2014. Exploration Geochemistry. Treat. Geochem. 623–650. Available from: https://doi.org/10.1016/b978-0-08-095975-7.01127-x.

Comrey, L.A., Lee, H.B., 1992. A First Course in Factor Analysis, second ed. Lawrence Erlbaum Associates, Hillside, NJ.

Darabi-Golestan, F., Hezarkhani, A., 2018. Evaluation of elemental mineralization rank using fractal and multivariate techniques and improving the performance by log-ratio transformation. J. Geochem. Explor. 189, 11–24.

Davis, J.C., 2002. Statistics and Data Analysis in Geology, third ed. John Wiley and Sons Inc, New York, 638 p.

Daya, A.A., Afzal, P., 2015. A comparative study of concentration-area (C-A) and spectrum-area (S-A) fractal models for separating geochemical anomalies in Shorabhaji region, NW Iran. Arab J. Geosci. 8, 8263–8275.

Delavar, S.T., Afzal, P., Borg, G., Rasa, I., Lotfi, M., Rashidnejad, O.N., 2012. Delineation of mineralization zones using concentration–volume fractal method in Pb–Zn carbonate hosted deposits. J. Geochem. Explor. 118, 98–110. Available from: https://doi.org/10.1016/j.gexplo.2012.05.003.

Dempster, M., Dunlop, P., Scheib, A., Cooper, M., 2013. Principal component analysis of the geochemistry of soil developed on till in Northern Ireland. J. Maps . Available from: https://doi.org/10.1080/17445647.2013.789414.

Drouin, M., 2012. Application of Factor Analysis in the Identification of a Geochemical Signature of Buried Kimberlites in Near-Surface Groundwaters in the Attawapiskat Area of the James Bay Lowlands of Northern Ontario, Canada. A thesis submitted to the Faculty of Graduate and Postdoctoral Studies in partial fulfillment of the requirements for the Master of Science Degree, in Earth Sciences. Ottawa-Carleton Geoscience Centre, Faculty of Science, University of Ottawa, Canada.

d'Uston, C., Gellert, R., 2006. Principal Component Analysis of Geochemical Data at Gusev Crater, XXXVII. Luna and Plannatery Sciences.

Egozcue, J.J., Pawlowsky-Glahn, V., Mateu-Figueraz, G., Barcelo-Vidal, C., 2003. Isometric log-ratio transformations for compositional data analysis. Math. Geosci. 35, 279–300.

Elachi, C., van Zyl, J., 2006. Introduction to the Physics and Techniques of Remote Sensing. John Wiley and Sons, Inc.

Fallara, F., Legault, M., Rabeau, O., 2006. 3-D integrated geological modeling in the Abitibi Subprovince (Québec, Canada): Techniques and applications. J. Explor. Mining Geol. 15, 27–43.

Feder, J., 1988. Fractals. Plenum Press, New York.

Fernandes, A.R., de Souza, E.S., de Souza Braz, A.M., Birani, S.M., Alleoni, L.R.F., 2018. Quality reference values and background concentrations of potentially toxic elements in soils from the Eastern Amazon, Brazil. J. Geochem. Explor. 190, 453–463.

Field, A., 2009. Discovering Statistics Using SPSS: Introducing Statistical Method, third ed. Sage Publications, Thousand Oaks, CA.

Filzmoser, P., Hron, K., 2008. Outlier detection for compositional data using robust methods. J. Math. Geosci. 40, 233–248.

Fletcher, W.K., 1981. Analytical Methods in Geochemical Prospecting. Elsevier, Amsterdam.

Florinsky, I.V., 2016. Predictive soil mapping, Chapter 11, 287-311pp. In: Igor, V. (Ed.), Floninsky: Digital Terrain Analysis in Soil Science and Geology, 2nd Edition Elsevier, 486p.

Ford, A., Blenkinsop, T.G., 2008. Combining fractal analysis of mineral deposit clustering with weights of evidence to evaluate patterns of mineralization: Application to copper deposits of the Mount Isa Inlier, NW Queensland, Australia. Ore Geol. Rev. 33, 435–450.

Ford, A., Peters, K.J., Partington, G.A., Blevin, P.L., Downes, P.M., Fitzherbert, J.A., et al., 2019. Translating expressions of intrusion-related mineral systems into mappable spatial proxies for mineral potential mapping: Case studies from the Southern New England Orogen. Austr. Ore Geol. Rev. 111 (102943), 1–18.

Fyzollahhi, N., Torshizian, H., Afzal, P., Jafari, M.R., 2018. Determination of lithium prospects using fractal modeling and staged factor analysis in Torud region, NE Iran. J. Geochem. Explor. 189, 2–10.

Gabr, S.S., Hassan, S.M., Sadek, M.F., 2015. Prospecting for new gold-bearing alteration zones at El-Hoteib area, South Eastern Desert, Egypt, using remote sensing data analysis. Ore Geol. Rev. 71, 1–13. Available from: https://doi.org/10.1016/j.oregeorev.2015.04.021.

Gandhi, S.M., Sarkar, B.C., 2016. Essentials of Mineral Exploration and Evaluation. Elsevier. 410p.

Gibson, P.J., 2000. Introductory Remote Sensing: Principles and Concepts. Routledge-Taylar and Francis Group.

Gill, S.B., Piercey, S.J., Layne, G.D., Piercey, G., 2019. Sulphur and lead isotope geochemistry of sulphide minerals from the Zn-Pb-Cu-Ag-Au Lemarchant volcanogenic massive sulphide (VMS) deposit, Newfoundland, Canada. Ore Geol. Rev. 104, 422–435. Available from: https://doi.org/10.1016/j.oregeorev.2018.11.008.

Gitelson, A.A., Buschmann, C., Lichtenthaler, H.K., 1998. Leaf chlorophyll fluorescence corrected for re-absorption by means of absorption and reflectance measurements. J. Plant. Physiol. 152 (2–3), 283–296. Available from: https://doi.org/10.1016/s0176-1617(98)80143-0.

Goldschmidt, V.M., 1937. The principles of distribution of chemical elements in minerals of rocks. Journal of Chemical Society of London 140, 655–673.

Goncalves, M.A., 2001. Characterization of geochemical distributions using multifractal models. Math. Geol. 33, 41–61.

Gonzalez, R.C., Woods, R.E., 2008. Digital Image Processing, third ed. Pearson Education, Inc, Upper Saddle River. New Jersey, ISBN: 978-0-13-168728-8.

González-Álvarez, I., Boni, M., Anand, R.R., 2016. Mineral exploration in regolith-dominated terrains: Global considerations and challenges. Ore Geol. Rev. 73, 375–379.

Gorsuch, R.L., 1983. Factor Analysis, second ed Lawrence Erlbaum Associates, Hillside, New Jersey.

Govett, G.J.S., 1978. Lithogeochemistry, letter to the editor. J. Geochem. Explor. 9, 109–110.

Govett, G.J.S., 1983. Handbook of Exploration Geochemistry, Rock Geochemistry in Mineral Exploration, third ed. Elsevier, Amsterdam, 461 p.

Govett, G.J.S., 2000. In: Hale, M. (Ed.), Geochemical Remote Sensing of Sub-Surface, Vol. 7. Elsevier Science B.V, Amsterdam, 573 p.

Govindaraju, K., 1994. Compilation of working values and sample description for 383 geostandards. Geostand. Newslett. 18, 1–158 (Special Issue).

Guadagnoli, E., Velicer, W.F., 1988. Relation to sample size to the stability of component patterns. Psychol. Bull. 103 (2), 265–275.

Haldar, S., 2018. Mineral Exploration: Principles and Applications, 2d ed. Elsevier, Amsterdam, 360p.

Harris, D.V.P., 1984. Mineral Resources Appraisal: Mineral Endowment, Resources, and Potential Supply: Concepts, Methods and Cases. Oxford University Press, London, 445 p.

Hassanpour, S., Afzal, P., 2013. Application of concentration–number (C–N) multifractal modeling for geochemical anomaly separation in Haftcheshmeh porphyry system, NW Iran. Arab J. Geosci. 6, 957–970.

Hill, R.E.T., Barnes, S.J., Gole, M.J., Dowling, S.E., 1995. The volcanology of komatiites as deduced from field relationships in the Norseman-Wiluna greenstone belt, Western Australia. Lithos 34, 159188.

Hosseini-Dinani, H., Aftabi, A., 2016. Vertical lithogeochemical halos and zoning vectors at Goushfil Zn–Pb deposit, Irankuh district, southwestern Isfahan, Iran: Implications for concealed ore exploration and genetic models. Ore Geol. Rev. 72, 1004–1021.

Hotelling, H., 1933. Analysis of a complex of statistical variables into principal components. J. Educ. Psychol. 24, 417–441. and 498-520.

Hronsky, J.M.A., Groves, D.I., 2008. Science of targeting: definition, strategies, targeting and performance measurement. Austr. J. Earth Sci. 55, 3–12.

Hu, Z., Gao, S., 2008. Upper crustal abundances of trace elements: a revision and update. Chem. Geol. 253 (3–4), 205–221. Available from: https://doi.org/10.1016/j.chemgeo.2008.05.010.

Hunt, G.R., 1980. Electromagnetic radiation—the communication link in remote sensing. In: Siegal, B.S., Gillespie, A.R. (Eds.), Remote Sensing in Geology. Wiley, New York.

Jarva, J., Tarvainen, T., Reinikainen, J., Eklund, M., 2010. TAPIR—Finnish national geochemical baseline database. Sci. Total Environ. 408, 4385–4395.

Jimenez-Espinosa, R., Sousa, A.J., Chica, O., 1993. Identification of geochemical anomalies using principal component analysis and factorial kriging analysisAmsterdam: Elsevier Science Publishers B.V J. Geochem. Explor. 46, 25–256.

Jolliffe, I.T., 2002. Principal Component Analysis, second ed. Springer–Verlag, New York.

Jöreskog, K.G., Klovan, J.E., Reyment, R.A., 1976. Geological Factor Analysis. Elsevier Scientific Publications, Amsterdam, 178 p.

Kaiser, H.F., 1958. The varimax criteria for analytical rotation in factor analysis. Psychometrika. 23, 187–200.

Kesler, S.E., 2000. Mineral Supply and Demand into the 21st Century. In Briskey and Schulz, 2000. Proceedings for a Workshop on Deposit Modeling, Mineral Resource Assessment, and their Role in Sustainable Development. U.S. Department of the Interior—U.S. Geological Survey.

Kesler, S.E., 2007. Mineral Supply and Demand into the 21st Century. In: Briskey, J., Schulz, K.(Eds.), Proceedings for a Workshop on Deposit Modeling, Mineral Resource Assessment, and their Role in Sustainable Development. United States Geological Survey: Virginia, 55–62.

Khalajmasoumi, M., Sadeghi, B., Carranza, E.J.M., MartiyaSadeghi, M., 2017. Geochemical anomaly recognition of rare earth elements using multi-fractal modeling correlated with geological features, Central Iran. J. Geochem. Explor. 181, 318–332.

Kleeman, A.W., 1967. Sampling error in the chemical analysis of rocks. J. Geol. Soc. Austr. 14 (1), 43–47.

Kline, P., 1994. An Easy Guide to Factor Analysis. Boutledge, New York.

Le Maitre, R.W., 1982. Numerical Petrology, 1s ed. Statistical Interpretation of Geochemical Data, 8. Earth and Planetary Sciences, eBook ISBN: 9781483289748.

Lebart, L., Morineau, A., Warwick, K., 1984. Multivariate Descriptive Statistical Analysis. John Wiley and Sons, New York.

Le Vaillant, M., Fiorentini, M.L., Barnes, S.J., Van Spronsen, J.W., 2018. Review of predictive and detective exploration tools for magmatic Ni-Cu-(PGE) deposits, with a focus on komatiite-related systems in Western Australia. In: Ore Deposits of Ultramafic-Mafic Magmas through Space and Time, 47–78. Available from: https://doi.org/10.1016/B978-0-12-811159-8.00003-2.

Li, C., Ma, T., Shi, J., 2003. Application of a fractal method relating concentrations and distances for separation of geochemical anomalies from background. J. Geochem. Explor. 77, 167–175.

Li, X., Yuan, F., Zhang, M., Jia, C., Jowitt, S.M., Ord, A., et al., 2015. Three-dimensional mineral prospectivity modeling for targeting of concealed mineralization within the Zhong gu iron or field, Ning wu Basin, China. Ore Geol. Rev. 71, 633–654.

Li, X., Yuan, F., Zhang, M., Jowitt, S.M., Ord, A., Zhou, T., et al., 2019. 3D computational simulation-based mineral prospectivity modeling for exploration for concealed Fe–Cu skarn-type mineralization within the Yueshan orefield, Anqing district, Anhui Province, China. Ore Geol. Rev. 105, 1–17.

Lillesand, T.M., Kiefer, R.W., 1994. Remote Sensing and Image Interpretation, third ed. John Wiley and Sons, Inc, Hoboken, 750 p.

Lin, S.-K., 2013. In: fifth ed. Campbell, J.B., Wynne, R.H. (Eds.), Introduction to Remote Sensing, 2011. The Guilford Press, 662p. Remote Sensing, 5(1), 282–283. https://doi.org/10.3390/rs5010282.

Lindsay, M.D., Betts, P.G., Ailleres, L., 2014. Data fusion and porphyry copper prospectivity models, southeastern Arizona. Ore Geol. Rev. 61, 120–140.

Liu, H., Huang, H., Li, G., Xue, W., Zhang, Z., Liu, B., et al., 2015. Factor analysis in geochemical survey of the Shangxu gold deposit. North. Tibet. Geol. China 42 (04), 1126–1136.

Macheyeki, A.S., 2003. Lithogeochemical Vectors Associated with the Kabanga Ni-Cu Sulphide Deposits North Western Tanzania: A Thesis Submitted in Partial Fulfillment of the Requirements for the Degree of Masters of Science in Geology in the Faculty of Science. University of the Western Cape, 129 p.

Macheyeki, A.S., 2011. Application of lithogeochemistry to exploration for Ni–Cu sulfide deposits in the Kabanga area, NW Tanzania. J. Afr. Earth Sci. 61, 62–81.

Maepa, F., Smith, R., 2017. Predictive mapping of the gold mineral potential in the Swayze Greentone Belt, ON, Canada. Society of Exploration Geophysicists. 2017 SEG International Exposition and Annual Meeting, 24–29 September, Houston, Texas. Document ID: SEG-2017-17660409.

Mandelbrot, B.B., 1983. The Fractal Geometry of Nature (updated and augmented edition). Freeman, New York, 495 p.

Mao, Z., Peng, S., Lai, J., Shao, Y., Yang, B., 2004. Fractal study of geochemical prospecting data in south area of Fenghuanshan copper deposit, Tongling Anhui. J. Earth Sci. Environ. 26 (4), 11–14.

Mao, X., Zou, Y., Cheng, J., Lai, J., Peng, S., 2011. Three-dimensional Visual Prediction of Concealed Orebodies. Central South University Press, Changsha (170 pp.) (in Chinese).

Mason, P.J., 2005. Remote Sensing | GIS. Encycl. Geol. 420–431. Available from: https://doi.org/10.1016/b0-12-369396-9/00114-3.

Maxwell, S.E., Delaney, H.D., 2000. Designing experiments and analyzing data: A model comparison perspective. Lawrence Erlbaum, Mahwah, NJ.

Maya, M., Musekiwa, C., Mthembi, P., Crowley, M., 2015. Remote sensing and geochemistry techniques for the assessment of coal mining pollution, Emalahleni (Witbank), Mpumalanga. South Afr. J. Geomat. 4 (2), 174–188.

McCuaig, T.C., Hronsky, J.M.A., 2000. The current status and future of the interface between the exploration industry and economic geology research. In: Hagemann, S.G., Brown, P.E. (Eds.), Gold in 2000, 13. Reviews in Economic Geology, 553–559 p.

McCuaig, T.C., Beresford, S., Hronsky, J., 2010. Translating the mineral systems approach into an effective exploration targeting system. Ore Geol. Rev. 38 (3), 128–138. Available from: https://doi.org/10.1016/j.oregeorev.2010.05.008.

McIlwaine, R., Cox, S.F., Doherty, R., 2016. Determining geochemical threshold values from the Tellus data sets: the examples of zinc and iodine. In: Young, M.E. (Ed.), Unearthed: Impacts of the Tellus Surveys of the North of Ireland. Royal Irish Academy, Dublin. Available from: https://doi.org/10.3318/978-1-908996-88-6.ch21.

McIlwaine, R., Doherty, R., Cox, S.F., Cave, M., 2017. The relationship between historical development and potentially toxic element concentrations in urban soils. Environ. Pollut. 220, 1036–1049. Available from: https://doi.org/10.1016/j.envpol.2016.11.040.

Meinert, L.D., 1997. Application of skarn deposit zonation models to mineral exploration. J. Explor. Min. Geol. 6, 185–208.

Moon, C.J., King, A., 2015. Geochemical Exploration for Tin Deposits: Application of Conventional and Novel Techniques in Cornwall, England. International Applied Geochemistry Symposium (Association of Applied Geochemists), Tucson, AZ USA.

MüCke, A., 1994. Chapter 11 Part I. Postdiagenetic ferruginization of sedimentary rocks (sandstones, oolitic ironstones, kaolins and bauxites)—including a comparative study of the reddening of red beds. Develop. Sedimentol. 361–395. Available from: https://doi.org/10.1016/s0070-4571(08)70444-8.

Muecke, G.K., Chatterjee, A.K., 1983. Lithogeochemistry as an indicator of uranium and tin mineralization, South Mountain Batholith, Nova Scotia, Canada. J. Geochem. Explor. 19 (1–3), 589–594.

Navidi, A., Ziaii, M., Afzal, P., Yasrebi, A.B., Wetherelt, A., Foster, P., 2014. Determination of chromites prospects using multifractal models and zonality index in the Parang 1:100000 Sheet, Iran. Univ. J. Geosci. 2 (4), 133–139.

Nielsen, S.H.H., Cunningham, F., Hay, R., Partington, G., Stokes, M., 2015. 3D prospectivity modelling of orogenic gold in the Marymia Inlier, western Australia. Ore Geol. Rev. 71, 578–591.

Ostadhosseini, A., Barati, M., Afzal, P., Lee, I., 2018. Prospecting polymetallic mineralization in Ardestan area, Central Iran, using fractal modeling and staged factor analysis. Geopersia 8 (2), 279–292.

Ott, N., Kollersberger, T., Tassara, A., 2006. GIS analyses and favorability mapping of optimized satellite data in northern Chile to improve exploration for copper mineral deposits. Geosphere 2 (4), 236–252.

O'Reilly, S.Y., Griffin, W.L., Belousova, E.A., 2004. Terrane Chron—Delivering a competitive edge in exploration. Predictive Mineral Discovery Under Cover. Symposium for the Society of Economic Geologists, 27 September–1 October 2004, Perth.

Pan, G.C., Harris, D.P., 2000. Information Synthesis for Mineral Exploration. Oxford University Press, New York.

Parasnis, D.S., 1996. Principles of Applied Geophysics, fifth ed Chapman and Hall, London.

Parsa, M., Maghsoudi, A., Yousefi, M., Sadeghi, M., 2017. Multifractal analysis of stream sediment geochemical data: Implications for hydrothermal nickel prospection in an arid terrain, eastern Iran. J. Geochem. Explor. 181, 305–317. Available from: https://doi.org/10.1016/j.gexplo.2016.11.013.

Parsapoor, A., Khalili, M., Maghami, M., 2017. Discrimination between mineralized and unmineralized alteration zones using primary geochemical haloes in the Darreh-Zar porphyry copper deposit in Kerman, southeastern Iran. J. Afr. Earth Sci. 132, 109–126.

Payne, C., Cunningham, F., Peters, K.J., Nielsen, S., Puccioni, E., Wildman, C., et al., 2015. From 2D to 3D: prospectivity modelling in the Taupo volcanic zone, New Zealand. Ore Geol. Rev. 71, 558–577.

Pearson, K., 1901. On lines and planes of closest fit to systems of points in space. Phil. Mag. 2 (11), 559–572.

Pereira, G.H., Renca, S., Saraiva, J., 2003. A case study on geochemical anomaly identification through principal components analysis supplementary projection. Appl. Geochem. 18 (1), 37–44. Available from: https://doi.org/10.1016/s0883-2927(02)00099-9.

Perrouty, S., Lindsay, M.D., Jessell, M.W., Aillères, L., Martin, R., Bourassa, Y., 2014. 3D modeling of the Ashanti Belt, southwest Ghana: evidence for a litho-stratigraphic control on gold occurrences within the Birimian Sefwi Group. Ore Geol. Rev. 63, 252–264. Available from: https://doi.org/10.1016/j.oregeorev.2014.05.011.

Porwal, A.K., Kreuzer, O.P., 2010. Introduction to the special issue: mineral prospectivity analysis and quantitative resource estimation. Ore Geol. Rev. 38 (3), 121–127.

Porwal, A., Gonzalez-Alvarez, I., Markwitz, V., McCuaig, T., Mamuse, A., 2010. Weights-of evidence and logistic regression modelling of magmatic nickel sulfide prospectivity in the Yilgarn Craton, Western Australia. Ore Geol. Rev. 38, 184–196.

Pounis, G., 2019. Statistical analysis of retrospective health and nutrition data. Anal. Nutr. Res. 103–144. Available from: https://doi.org/10.1016/b978-0-12-814556-2.00005-1.

Pour, A.B., Hashim, M., 2012. The application of ASTER remote sensing data to porphyry copper and epithermal gold deposits. Ore Geol. Rev. 44, 1–9.

Ramanaidou, E., Wells, M., Lau, I., Laukamp, C., 2015. Characterization of iron ore by visible and infrared reflectance and, Raman spectroscopies. Iron Ore, 191–228. https://doi.org/10.1016/B978-1-78242-156-6.00006-X.

Regmi, N.R., Rasmussen, C., 2018. Predictive mapping of soil-landscape relationships in the arid Southwest United States. CATENA 165, 473–486.

Reimann, C., Filzmoser, P., Garrett, R.G., 2002. Factor analysis applied to regional geochemical data: problems and possibilities. J. Appl. Geochem. 17, 185–206.

Reimann, C., Filzmoser, P., Garrett, R.G., Dutter, R., 2008. Statistical Data Analysis Explained, 1s ed John Wiley and Sons, Chichester, UK.

Reimann, C., Schilling, J., Roberts, D., Fabian, K., 2015. A regional-scale geochemical survey of soil O and C horizon samples in Nord-Trøndelag, Central Norway: geology and mineral potential. Appl. Geochem. 61, 192–205. Available from: https://doi.org/10.1016/j.apgeochem.2015.05.019.

Rekhibi, S., Wadi, M., Said, A., 2015. Remote Sensing and GIS Techniques for Gold Exploration. Int'l Conf. on Advances in Science, Engg., Technology and Natural Resources (ICASETNR-15) Aug. 27–28, Kota Kinabalu (Malaysia), 51–56. https://doi.org/10.15242/IICBE.C0815061.

Reyment, A., Jöreskog, K.G., 1996. Applied Factor Analysis in the Natural Sciences. University press, Cambridge, UK.

Richards, J.P., 2011. High Sr/Y arc magmas and porphyry Cu ± Mo ± Au deposits. Econ. Geol. 106, 1075–1081.

Rohrlach, B.D., Loucks, R.R., 2005. Multi-million-year cyclic ramp-up of volatiles in a lower crustal magma reservoir trapped below the Tampakan copper–gold deposit by Mio-

Pliocene crustal compression in the southern Philippines. In: Porter, T.M. (Ed.), Super Porphyry Copper and Au Deposits: A Global Perspective. PGC Publishing, Adelaide.

Rollinson, H.R., 1993. Using Geochemical Data: Evaluation, Presentation, Interpretation. Longman Group Ltd, United Kingdom, pp. 150–170 (Geochemical series).

Rouvray, D.H., 1996. The geometry of nature. Endeavour 20 (2), 79–85. Available from: https://doi.org/10.1016/0160-9327(96)10010-7.

Roy, R., Cassard, D., Cobbold, P.R., Rossello, E.A., Billa, M., Bailly, L., et al., 2006. Predictive mapping for copper–gold magmatic-hydrothermal systems in NW Argentina: Use of a regional-scale GIS, application of an expert-guided data-driven approach, and comparison with results from a continental-scale GIS. Ore Geol. Rev. 29 (3–4), 260–286. Available from: https://doi.org/10.1016/j.oregeorev.2005.10.002.

RSC, Mining and Mineral Exploration Co. 2015. Detailed and systematic mineral exploration programme in the Mwogo and Kirehe regions in Rwanda. Tender Number 019/RNRA/13/14.

Rugless, C.S., 1983. Lithogeochemistry of Wainaleka Cu-Zn volcanogenic deposit, Viti Levu, Fiji, and possible applications for exploration in tropical terrains. J. Geochem. Explor. 19, 563–586.

Sabins, F.F., 1983. Geologic interpretation of Space Shuttle images of Indonesia. Am. Assoc. Petrol. Assoc. Bull. 67, 2076–2099.

Sabins, F.F., 1987. Remote Sensing—Principles and Interpretation, second ed. W.H. Freeman, New York, 449 p.

Sabins, F.F., 1997. Remote Sensing—Principles and Interpretation, third ed. W.H. Freeman, New York, 494 p.

Sabins, F.F., 1999. Remote sensing for mineral exploration. Ore Geol. Rev. 14, 157–183.

Sadeghi, M., Morris, G.A., Carranza, E.J.M., Ladenberger, A., Andersson, M., 2013. Rare earth element distribution and mineralization in Sweden: an application of principal component analysis to FOREGS soil geochemistry. J. Geochem. Explor. 133, 160–175.

Sadiya, T.B., Ibrahim, O., Asma, T.F., Mamfe, V., Nsofor, C.J., Oyewmi, A.S., et al., 2014. Mineral detection and mapping using band ratioing and Crosta technique in Bwari Area Council, Abuja Nigeria. Int. J. Scient. Eng. Res. 5 (12), 1100–1108.

Safari, S., Ziaii, M., Ghoorchi, M., Sadeghi, M., 2018. Application of concentration gradient coefficients in mining geochemistry: A comparison of copper mineralization in Iran and Canada. J. Mining Environ. 9 (1), 277–292.

Sahoo, P.K., Guimar, T.F., Wal, P., Filho, M.S., da Silva, M.S., Maurity, C.W., et al., 2016. Geochemistry of upland lacustrine sediments from Serra dos Carajàs, Southeastern Amazon, Brazil: implications for catchment weathering, provenance, and sedimentary processes. J. South Am. Earth Sci. 72, 178–190. Available from: https://doi.org/10.1016/j.jsames.2016.09.003.

Sakar, M., Chandan, H.R., Shwetharani, R., 2019. Graphene paper-based electrochemical sensors for biomolecules. Graph. Based Electrochem. Sens. Biomol. 297–320. Available from: https://doi.org/10.1016/b978-0-12-815394-9.00012.

Sara, C., 2011. Source identification of environmental pollutants using chemical analysis and Positive Matrix Factorization. PhD Thesis. Facoltà di Scienze Matematiche, Fisiche e Naturali Dipartimento di Scienze della Terra "Ardito Desio". Università Degli Studi Di Milano.

Schetselaar, E.M., Harris, J.R., Lynds, T., de Kemp, E.A., 2008. Remote predictive mapping 1. Remote predictive mapping (RPM): a strategy for geological mapping of Canada's North. J. Geol. Soc. Canada 34, 3–4.

Scott, K.M., 2005. Rutile geochemistry as a guide to porphyry Cu-Au mineralization, Northparkes, New South Wales, Australia. Geochem-Explor Env A 5 (3), 247–253.

Seneshen, D.M., 2015. Geochemical Exploration for Conventional and Unconventional Petroleum Resources. International Applied Geochemistry Symposium (Association of Applied Geochemists), Tucson, AZ, USA.

Shi, C., Wang, C., 1995. Regional geochemical secondary negative anomalies and their significance. J. Geochem. Explor. 55 (1–3), 11–23. Available from: https://doi.org/10.1016/0375-6742(95)00033-x.

Shi, Y., Ji, H., Lu, J., Ma, L., Duan, G., 2004. Factor analysis method and application of stream sediment geochemical partition. Geol. Prosp. 05, 73–76.

Shuguang, Z., Kefa, Z., Yao, C., Jinlin, W., Jianli, D., 2015. Exploratory data analysis and singularity mapping in geochemical anomaly identification in Karamay, Xinjiang, China. J. Geochem. Explor. 154, 171–179.

Siegal, B.S., Gillespie, A.R., 1980. Remote Sensing in Geology. John Wiley and Sons, Inc, USA, 702 p.

Sinclair, A.J., 1974. Selection of threshold values in geochemical data using probability graphs. J. Geochem. Explor. 3, 129–149.

Smith, B.H., 1984. Geochemical Exploration for Nickel Sulphides in Lateritic Terrain in Western Australia. Western Mining Corporation, Limited, Perth, Western Australia, pp. 40–41.

Sochevanov, N.N., 1961. Method of sampling of underground workings and surface in search of ore bodies and deposits on the primary dispersion halos. B. Sat: Problems. Techniques for testing of ore deposits in the exploration and exploitation. Gosgeoltekhizdat (in Russian). on gold occurrences within the Birimian Sefwi Group. Ore Geol. Rev. 63, 252–264.

Soltani, F., Afzal, P., Asghari, O., 2014. Delineation of alteration zones based on Sequential Gaussian Simulation and concentration–volume fractal modeling in the hypogene zone of Sungun copper deposit, NW Iran. J. Geochem. Explor. 140, 64–76. Available from: https://doi.org/10.1016/j.gexplo.2014.02.007.

Sprague, K., de Kemp, E., Wong, W., McGaughey, J., Perron, G., Barrie, T., 2006. Spatial targeting using queries in a 3-D GIS environment with application to mineral exploration. J. Comput. Geosci. 32, 396–418.

Stanley, H.E., Meakin, P., 1988. Multifractal phenomena in pysics and chemistry. Nature 333, 405–409.

Tabachnick, B.G., Fidell, L.S., 2007. Using Multivariate Statistics, fifth ed. Allyn and Bacon, Boston, MA.

Telford, W.M., Geldart, L.T., Sherrif, R.E., 1990. Applied Geophysics, second ed. Cambridge University Press.

Toth, C., Jóźków, G., 2016. Remote sensing platforms and sensors: a survey. ISPRS J. Photogrammetr. Rem. Sens. 115, 22–36. Available from: https://doi.org/10.1016/j.isprsjprs.2015.10.004.

Turcotte, D.L., 1997. Fractals and Chaos in Geology and Geophysics, second ed Cambridge University Press, United Kingdom, 398 p.

Wang, G., Huang, L., 2012. 3D geological modeling for mineral resource assessment of the Tongshan Cu deposit, Heilongjiang Province, China. J. Geosci. Front. 3, 483–491.

Wang, X., Wen, X., Rong, Y., Liu, Z., Sun, B., Zhao, S., et al., 2007. Vertical variations and dispersion of elements in arid desert regolith: a case study from the Jinwozi gold deposit, northwestern China. Geochem. Explor. Environ. Anal. 7, 163–171.

Wang, G., Zhang, S., Yan, C., Song, Y., Sun, Y., Li, D., et al., 2011. Mineral potential targeting and resource assessment based on 3D geological modeling in Luanchuan region, China. J. Comput. Geosci. 37, 1976–1988.

Wash, M.J., 2003. Remote Sensing. Department of the Army. US Corps of Engineering, Manual No. 1110-2-2907.

Wellington, T.-A.A., Mason, T.E., 2014. The effects of population growth and advancements in technology on global mineral supply. Resour. Policy 42, 73–82. Available from: https://doi.org/10.1016/j.resourpol.2014.10.006.

White, N.C., Hedenquist, J.W., 1990. Epithermal environments and styles of mineralization: variations and their causes, and guidelines for exploration. J. Geochem. Explor. 36, 445–474.

Wilkinson, J.J., 2014. Sediment-hosted zinc–lead mineralization. Treat. Geochem. 219–249.

Xiang, Z., Gu, X., Wang, E., Wang, X., Zhang, Y., Wang, Y., 2018. Delineation of deep prospecting targets by combining factor and fractal analysis in the Kekeshala skarn Cu deposit, NW China. J. Geochem. Explor. 198, 71–81.

Xiao, K., Li, N., Sun, L., Zou, W., Li, Y., 2012. Large scale 3D mineral prediction methods and channels based on 3D information technology. J. Geol. 36, 229–236 (In Chinese with English abstract).

Xiao, K., Li, N., Porwal, A., Holden, E.J., Bagas, L., Lu, Y., 2015. GIS-based 3D prospectivity mapping: a case study of Jiama copper-polymetallic deposit in Tibet, China. Ore Geol. Rev. 71, 357–367.

Xiao, K., Xing, S., Bagas, L., Sun, L., Li, N., Yin, J., et al., 2017. The China national mineral assessment initiative. Ore Geol. Rev. 91, 1084–1093.

Xiong, Y., Zuo, R., Carranza, E.J.M., 2018. Mapping mineral prospectivity through big data analytics and a deep learning algorithm. Ore Geol. Rev. 102, 811–817. Available from: https://doi.org/10.1016/j.oregeorev.2018.10.006.

Yinggui, L., Cheng, H., Xuedong, Y., Waisheng, X., 1995. Geochemical Exploration for concealed Ni-Cu deposits. J. Geochem. Explor. 55, 309–320.

Yong, A.G., Pearce, S., 2013. A beginner's guide to factor analysis: focusing on exploratory factor analysis. Tutor. Quantit. Methods Psychol. 9 (2), 79–94.

Yousefi, M., 2017. Recognition of an enhanced multi-element geochemical signature of porphyry copper deposits for vectoring into mineralized zones and delimiting exploration targets in Jiroft area, SE Iran. Ore Geol. Rev. 83, 200–214.

Yousefi, M., Carranza, E.J.M., 2015. Prediction-area (P-A) plot and C-A fractal analysis to classify and evaluate evidential maps for mineral prospectivity modeling. Comput. Geosci. 79, 69–81.

Yousefi, M., Kamkar-Rouhani, A., Carranza, E.J.M., 2014. Application of staged factor analysis and logistic function to create a fuzzy stream sediment geochemical evidence layer for mineral prospectivity mapping. Geochem. Explor. Environ. Anal. 14, 45–58.

Yuan, F., Li, X., Zhang, M., Jowitt, S.M., Jia, C., Zheng, T., et al., 2014a. Three-dimensional weights of evidence-based prospectivity modeling: a case study of the Baixiangshan mining area, Ningwu Basin, Middle and Lower Yangtze Metallogenic Belt, China. J. Geochem. Explor. 145, 82–97.

Yuan, F., Li, X., Zhang, M., Zhou, T., Gao, D., Hong, D., et al., 2014b. Three dimension prospectivity modelling based on integrated geo information for prediction of buried orebodies. J. Acta Geol. Sinica 88, 630–643 (In Chinese with English abstract).

Yuan, F., Li, X., Zhou, T., Deng, Y., Zhang, D., Xu, C., et al., 2015. Multifractal modeling-based mapping and identification of geochemical anomalies associated with Cu and Au mineralization in the NW Junggar area of northern Xinjiang Province, China. J. Geochem. Explor. 154, 252–264.

Yuan, F., Li, X., Zhang, M., Jia, C., Hu, X., 2018. Research progress of 3D prospectivity modeling. Gansu Geo-eng. 27, 32–36 (In Chinese with English abstract).

Zakaria, N., 2014. Body shape analysis and identification of key dimensions for apparel sizing systems. Anthropomet., Apparel Sizing Design 95–119. Available from: https://doi.org/10.1533/9780857096890.1.95.

Ziaii, M., 2008. The application of concentration gradient of zonality coefficient for indicator elements as a criterion in mining geochemistry. Goldschmidt Conference.

Ziaii, M., Safari, S., Timkin, T., Voroshilov, V., Yakich, T., 2019. Identification of geochemical anomalies of the porphyry–Cu deposits using concentration gradient modelling: a case study, Jebal-Barez area, Iran. J. Geochem. Explor. 199, 16–30.

Zuo, R., Wang, J., 2016. Fractal/multifractal modeling of geochemical data: a review. J. Geochem. Explor. 164, 33–41.
Zuo, R., Xiong, Y., 2018. Big data analytics of identifying geochemical anomalies supported by machine learning methods. Nat. Resour. Res. 27, 5–13.
Zuo, R., Cheng, Q., Agterberg, F.P., Xia, Q., 2009. Application of singularity mapping technique to identify local anomalies using stream sediment geochemical data, a case study from Gangdese, Tibet, Western China. J. Geochem. Explor. 101, 225–235.
Zuo, R., Carranza, E.J.M., Cheng, Q., 2012. Fractal/multifractal modelling of geochemical exploration data. J. Geochem. Explor. 122, 1–3.
Zuo, R., Wang, J., Chen, G., Yang, M., 2015. Identification of weak anomalies: a multifractal perspective. J. Geochem. Explor. 148, 12–24.

Further reading

Aggarwal, S., 2001. Principles of Remote sensing. In: Sivakumar, M.V.K., Roy, P.S., Harmsen, K., Saha, S.K. (Eds.), Satellite Remote Sensing and GIS Applications in Agricultural Meteorology, 2004. World Meteorological Organization, Switzerland., AGM-8. WMO/TD No. 1182.
Anders, E., Ebihara, M., 1982. Solar-system abundances of the elements. Geochim. Cosmochim. Acta 46 (11), 2363–2380.
Cao, M., Lu, L., 2015. Application of the multivariate canonical trend surface method to the identification of geochemical combination anomalies. J. Geochem. Explor. 153, 1–10.
Chen, G.-N., Grapes, R., 2007. Granite Genesis: In Situ Melting and Crustal Evolution. Springer, Dordrecht: the Netherlands, 278 p.
Cheng, Q., 2005. A new model for incorporating spatial association and singularity in interpolation of exploratory data. In: Leuanthong, D., Deutsch, C.V. (Eds.), Geostatistics Banff, 2004. Springer, Dordrecht: Netherlands, pp. 1017–1025.
Cheng, Q., 2012. Singularity theory and methods for mapping geochemical anomalies caused by buried sources and for predicting undiscovered mineral deposits in covered areas. J. Geochem. Explor. 122, 55–70.
Cohen, D.R., Booth, G.W., Govett, G.J.S., Beck, R.W., 1992. A statistical and analytical approach to the sizing of particulate gold. J. Geochem. Explor. 43 (1), 1–23. Available from: https://doi.org/10.1016/0375-6742(92)90021-y.
Emery, W., Camps, A., 2017. Land Applications. Introduction to Satellite Remote Sensing, 701–766.
Filzmoser, P., Hron, K., Reimann, C., 2009. Univariate statistical analysis of environmental (compositional) data—problems and possibilities. Sci. Total Environ. 407, 6100–6108.
Filzmoser, P., Hron, K., Reimann, C., 2010. The bivariate statistical analysis of environmental (compositional) data. Sci. Total Environ. 408, 4230–4238.
Goodfellow, W.D., McCutcheon, S.R., Peter, J.M., 2003. Massive sulfide deposits of the Bathurst Mining Camp, New Brunswick, and Northern Maine: introduction and summary of findings. Soc. Explor. Geophysicists Geophys. Monograph Series 11, 1–16. Available from: https://doi.org/10.5382/Mono.11.
Hawkes, H.E., 1957. Principles of Geochemical Prospecting. Contribution to geochemical prospecting for minerals. US Government Printing, pp. 225–355.
Herbert, S., Woldai, T., Carranza, E.J.M., van Ruitenbeek, F.J.A., 2014. Predictive mapping of prospectivity for orogenic gold in Uganda. J. Afr. Earth Sci. 99, 666–693.
http://www.gdmc.nl/oosterom/PoRSHyperlinked.pdf.
http://www.physlink.com/education/askexperts/ae366.cfm.
https://en.wikipedia.org/wiki/Eigenvalues_and_eigenvectors (09 September, 2018).
https://www.itc.nl/library/papers_2009/general/principlesremotesensing.pdf.

https://www.nrcan.gc.ca/node/14623#answer.
https://www.researchgate.net/post/How_to_interpret_factor_scores_from_Exploratory_Factor_Analysis (01 Sept. 2018).
Japan Oil, Gas and Metals National Corporation (JOGMEC), 2010. Introduction to Remote Sensing. Botswana Geologic Remote Sensing Centre. Training materials.
Large, D.E., 1980, Geologic parameters associated with sediment-hosted, submarine exhalative Pb-Zn deposits: An empirical model for mineral exploration, in Stratiform Cu-Pb-Zn deposits: Geologisches Jahrbuch series D40, 59–129.
Large, D.E., 1981. Sediment-hosted submarine exhalative lead-zinc deposits--A review of their geological characteristics and genesis. In: Wolf, K.H. (Ed.), Handbook of strata-bound and stratiform ore deposits, 9. Elsevier, pp. 469–508.
Large, D.E., 1983. Sediment-hosted massive sulfide lead-zinc deposits: An empirical model. In: Sangster, D.F. (Ed.), Sediment-hosted stratiform lead-zinc deposits, 8. Mineralogical Association of Canada Short Course Handbook, pp. 1–30.
Le Bas, M.J., Streckeisen, A.L., 1991. The IUGS systematics of igneous rocks (On behalf of the International Union of Geological Sciences, Subcommission on the Systematics of Igneous Rocks). J. Geol. Soc. Lond. 148, 825–833.
Le Maitre, 2002. Igneous Rocks: A Classification and Glossary of Terms. Cambridge University Press.
Lewis, R. 2000. Fractals. In Your Future. Chapter 1. Ontario.
Li, Y., Cheng, H., Yu, X., Xu, W., 1995. Geochemical exploration for concealed nickel-copper deposits. J. Geochem. Explor. 55 (1–3), 309–320. Available from: https://doi.org/10.1016/0375-6742(94)00065-4.
Maier, W.D., Barnes, S.J., de Waal, S.A., 1998. Exploration for magmatic Ni-Cu-PGE sulphide deposits: a review of recent advances in the use of geochemical tools, and their application to some South African Ores. South Afr. J. Geol. 101 (3), 237–253.
Meija, J., Coplen, T.B., Berglund, M., Brand, W.A., Bièvre, P.D., Gröning, M., et al., 2013. Atomic weights of the elements (IUPAC Technical Report). Pure Appl. Chem. Available online 2016 (www.iupac.org).
Peach, C.L., Mathez, E.A., 1996. Constraints on the formation of Platinum-Group Element Deposits in Igneous Rocks. J. Econ. Geol. 91, 439–450.
Pichard, C., Amossé, J., Piboule, M., Giovenazzo, D., 1995. Physical and Chemical Constraints on Platinum-Group Element Behavior during Crystallization of a Balsaltic Komatiite Liquid: Examples of the Proterozoic Delta Sill, New Quebec, Canada. J. Econ. Geol. 90, 2287–2303.
Ridley, J., 2013. Ore deposit Geology. Cambridge University Press, UK.
Turner, M.J., 2000. Modeling Nature with Fractals. Leicester.
U.S. Bureau of Mines, 1956. Mineral facts and problems. U.S. Bureau of Mines Bulletins, 556 (1042 p.) and 671 (1060 p.).
U.S. Bureau of Mines, 1993. Metal Prices in the United States through 1991. U.S. Bureau of Mines, Washington, D.C, 201 p.
Van Spronsen, J.W., 1969. The Periodic System of Chemical Elements. Elsevier, Amsterdam, pp. 315–316.
www.msscience.com (http://www.clark-shawnee.k12.oh.us/userfiles/36/Classes/1702/chap15.pdf?id=2787).

CHAPTER

4

Application of nonconventional mineral exploration techniques: case studies

SUBCHAPTER

4.1

In Africa

SUBCHAPTER

4.1.1

The use of lithogeochemical vectors in the Kabanga Nickel Project, Tanzania

Athanas S. Macheyeki

Earth Sciences Institute of Shinyanga, Shinyanga, United Republic of Tanzania

Applied Geochemistry
DOI: https://doi.org/10.1016/B978-0-12-819495-9.00004-9

© 2020 Elsevier Inc. All rights reserved.

4.1.1.1 Geology and mineralization

The Kabanga Ni–Cu deposit in Tanzania is located about 200 km SW of Lake Victoria in Tanzania. It is within the Mesoproterozoic Karangwe–Ankolean tectonic domain that is mainly dominated by mildly metamorphosed sediments. The metasedimentary rocks were intruded by series of mafic–ultramafic rocks in Mesoproterozoic era and by syn-tectonic and post-tectonic granites (Ikingura et al., 1992; Tack et al., 1994; Macheyeki, 2011) (Fig. 4.1.1).

The Kabanga Ni–Cu sulfide deposits (Kabanga main and Kabanga north ore deposits, Fig. 4.1.1), are characterized by both the Kabanga main and Kabanga north mafic–ultramafic rocks are crudely layered and differentiated, with cumulate olivine dominating and the western part and gabbroic rocks dominating the eastern part, such that they young to the east (Evans et al., 2000). Three main types of mineralization styles characterize these deposits: (1) peripheral veins in the country-rock metapelites, that is, detached ore bodies; (2) contact type massive sulfide mineralization, that is, ore bodies next to ultramafic rocks; and (3) disseminated and vein mineralization associated with the layered silicate rocks of the central zone of the intrusion, that is ore bodies within ultramafic rocks (Macheyeki, 2003, 2011).

4.1.1.2 Methodology and results

Rock samples collected on either type of the mineralization styles stated in section 4.1.1.1 were crushed and pulverized using an agate mill, into powder below 75 μm prior to analysis. Soil samples collected on profiles as illustrated in Fig. 4.1.2 were dried, sieved, and analyzed. Procedures for preparation and analysis of both rock and soil samples are in accordance with Hall and Bonham-Carter (1988) and Chao and Sanzolone (1992) and as described by Macheyeki (2003, 2011).

Statistical computation of geochemical data (e.g., principal component analysis) as narrated in section 3.2.2.1 of this book indicate about four main clusters of element grouping within ore bodies and host/country rock: (1) Ni–Zn–Pb–Cu, and partly Zr representing ore elements; (2) Mo–Cr–V representing elements hosted by magnetite/chromites and oxides; (3) Ba–Sr–Rb–Y representing elements hosted by feldspars; and (4) U–Th–Ce–La–Ga–Nb as a cluster of elements from accessory minerals (Fig. 4.1.3).

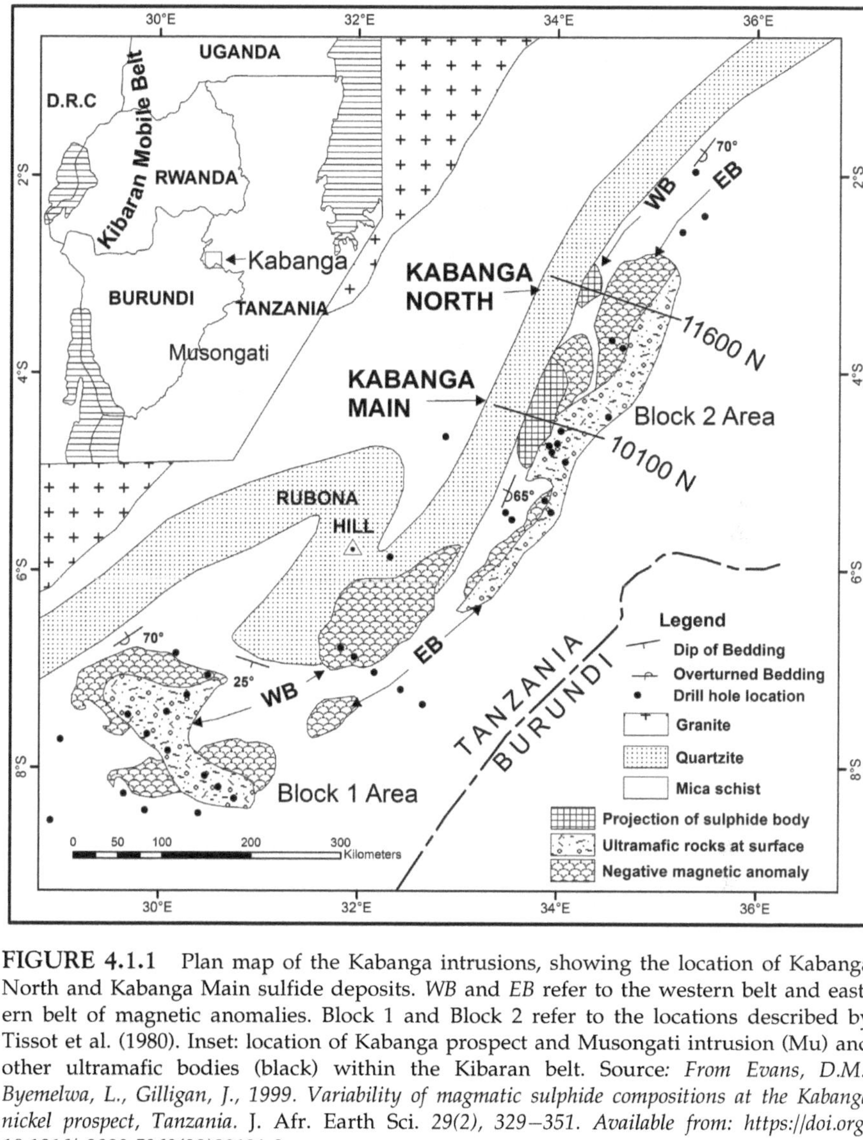

FIGURE 4.1.1 Plan map of the Kabanga intrusions, showing the location of Kabanga North and Kabanga Main sulfide deposits. *WB* and *EB* refer to the western belt and eastern belt of magnetic anomalies. Block 1 and Block 2 refer to the locations described by Tissot et al. (1980). Inset: location of Kabanga prospect and Musongati intrusion (Mu) and other ultramafic bodies (black) within the Kibaran belt. Source: *From Evans, D.M., Byemelwa, L., Gilligan, J., 1999. Variability of magmatic sulphide compositions at the Kabanga nickel prospect, Tanzania.* J. Afr. Earth Sci. *29(2), 329–351. Available from: https://doi.org/10.1016/s0899-5362(99)00101-3.*

Generally, lithogeochemical vectors related to Ni-sulfide mineralization can be derived from element ratios across the first two groups (Ni–Zn–Pb–Cu and Mo–Cr–V) and may reflect the degree of segregation of sulfide elements from silicate (e.g., Brand, 1999) such as Ni/Mo, Ni/Cr, Ni, V, Zn/Mo, Zn/Cr, Zn/V, Pb/Mo, Pb/Cr, Pb/V, Cu/Mo,

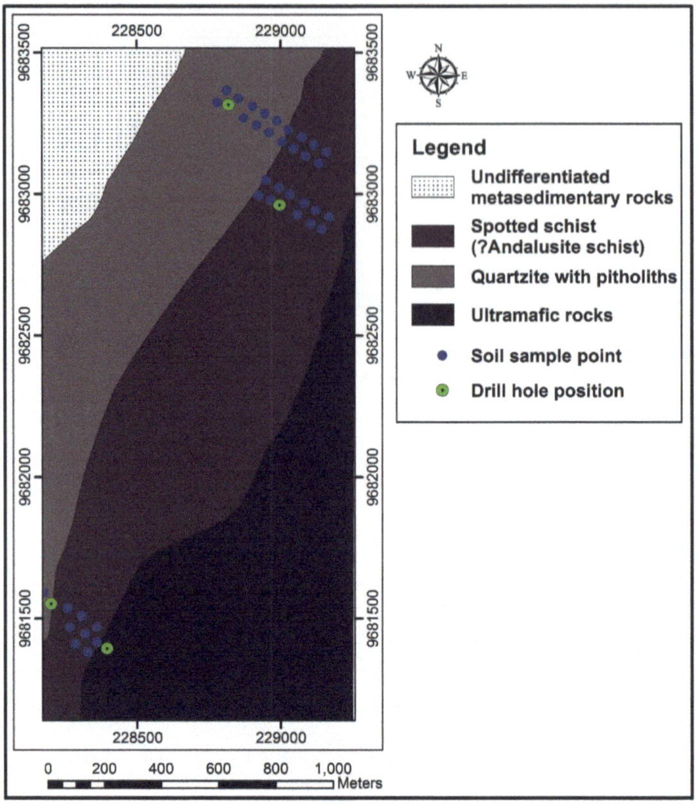

FIGURE 4.1.2 The sketch of Kabanga Main Ni–Cu sulfide deposit in plan view and the approximate locations of the sampled drill holes KN9869 and KN9873, over which, soil sample profiles were established. Source: *After Macheyeki, A.S., 2003. Lithogeochemical vectors associated with the Kabanga Ni–Cu sulfide deposits, NW Tanzania. Unpublished M. Sc. Thesis, The University of the Western Cape, Cape Town, 117 p.*

Cu/Cr, and Cu/V. When elements (including platinum group elements (PGE)) are plotted against the different ores (Fig. 4.1.4), it is clear from both ore intersections that Pt and Pd show relatively more consistent patterns than those of Ni and Cu. In both ores, V and PGE (particularly Pd) show characteristic trends in such a way that Pd generally increases with decrease in V and vice versa. Chromium also shows similar patterns with V. The pattern of PGE, Ni, and Cu is also important. Although both Ni and Cu have higher values in massive ores, they increase or decrease in relatively opposite directions, in such a way that, highest values of Ni correspond with lowest values of Cu. On the other hand, Pt and Pd contents are in resonance with those of Cu. At

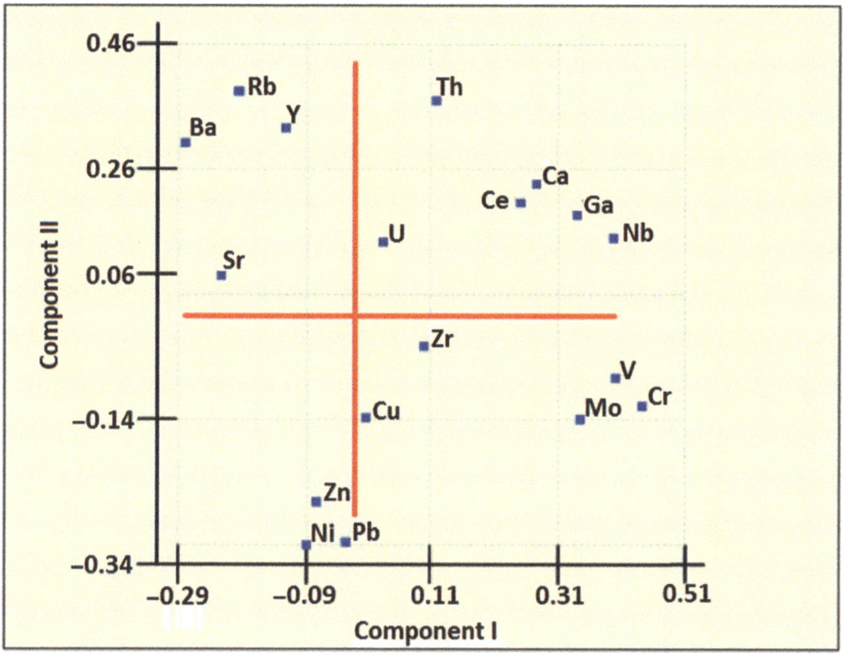

FIGURE 4.1.3 The plot of principal component I against II for the trace elements obtained from Kabanga drill holes. Source: *From Macheyeki, A.S., 2011. Application of litho-geochemistry to exploration for Ni–Cu sulfide deposits in the Kabanga area, NW Tanzania.* J. Afr. Earth Sci. *61, 62-81.*

least two conclusions can be drawn from these patterns: (1) PPGE and Cu show similar patterns probably because they are all relatively more strongly chalcophile than Ni and (2) Pd/V, Pd/Cr, Pt/V, Pt/Cr, Cu/V, Cu/Cr, Ni/Pt, Ni/Pd, and Ni/Cu ratios may be useful to vector toward Ni–Cu mineralization owing to their opposite trends within the ores.

Detailed petrography work on contact ores/ores located next to ultramafic rocks (mainly Kabanga Main), shows that Ni content slightly increases with increase in pyrrhotite/pentlandite ratios (Fig. 4.1.4) and increase in niccolite content and so highlights that, secondary sulfides other than pentlandite could have been responsible for Ni–Cu mineralization in these ores.

The most successful lithogeochemical vectors or element ratios derived for the Kabanga Ni–Cu sulfide deposits include (Pd/V) $\times$ 1000, and [(Pd/V)/(Cu/Cr)] $\times$ 100; Pd in ppb, whereas other elements are in ppm.

Population trends in the lower and upper cutoff values for the vectors were determined from histograms and cumulative frequency,

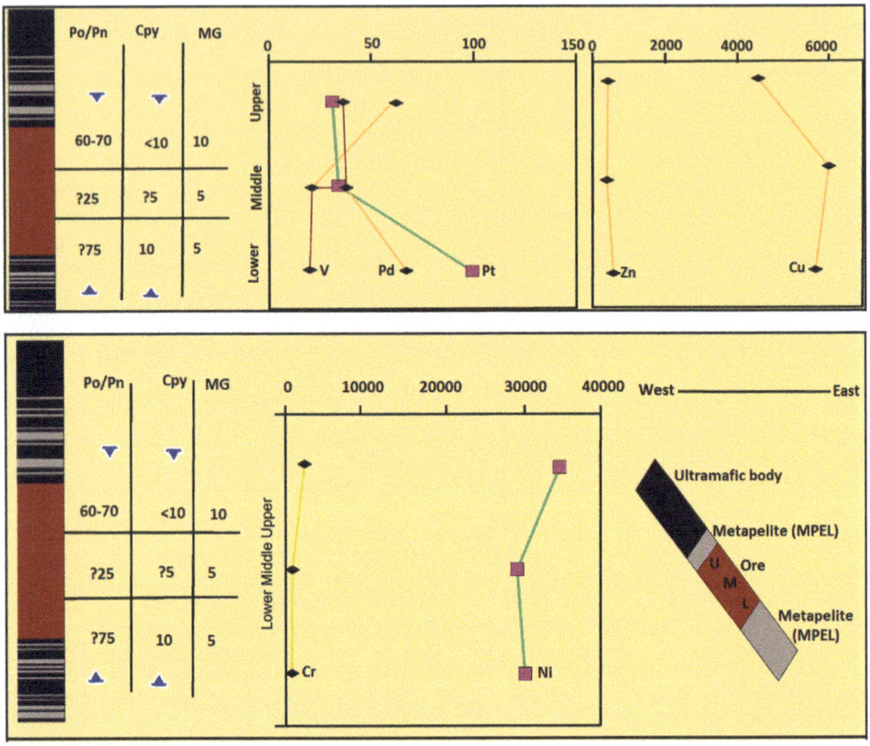

FIGURE 4.1.4 Element variability in detached ores along with petrography (lithologic unit) from Kabanga. Po/Pn, Cpy, and MG, respectively represent pyrrhotite/pentlandite ratio, chalcopyrite (estimated in %), and magnetite or chromite (estimated in %). *U*, *M*, and *L* stand for upper, middle, and lower intersection, respectively. An *arrow* points to decreasing values in a particular column. All element concentrations are in ppm, except for Pt and Pd that are in ppb. Source: *From Macheyeki, A.S., 2011. Application of lithogeochemistry to exploration for Ni–Cu sulfide deposits in the Kabanga area, NW Tanzania.* J. Afr. Earth Sci. *61, 62-81.*

respectively. An example on how to obtain the quantifiable upper and lower limits of the vectors is given for the (Pd/V) × 1000 as presented in Fig. 4.1.5A and B.

From Fig. 4.1.5A and B, it can be derived that, the lower limit cutoff value, for (Pd/V) × 100 is 195, whereas for upper limit cutoff values, it is about 1000. Using similar approaches, the upper and lower limits for [(Pd/V)/(Cu/Cr)] × 100 is 180 and 600, respectively (Macheyeki, 2003).

These lithogeochemical vectors were applied in a new area 50 km NE of Kabanga Ni–Cu sulfide deposits (Luhuma prospect) and corresponded with massive sulfides intersections at depth (Fig. 4.1.6).

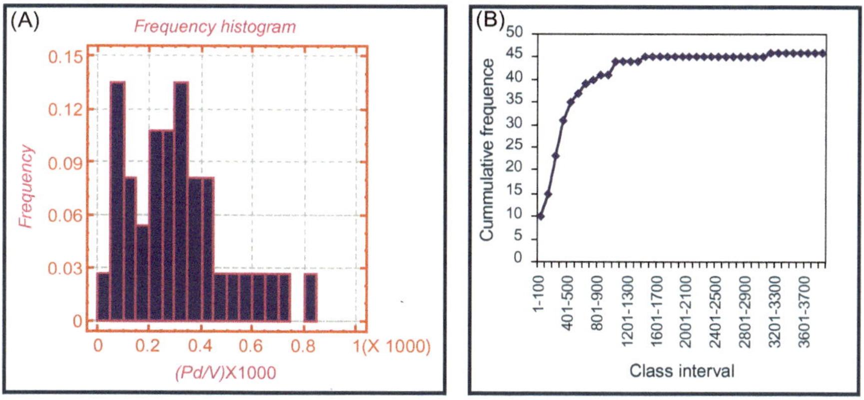

FIGURE 4.1.5 (A) Histogram of (Pd/V) × 1000 based on 20 classes. First major break occurs at 95. (B) Cumulative frequency against class intervals (40 classes) for (Pd/V) × 1000. Major break occurs at 901–1000.

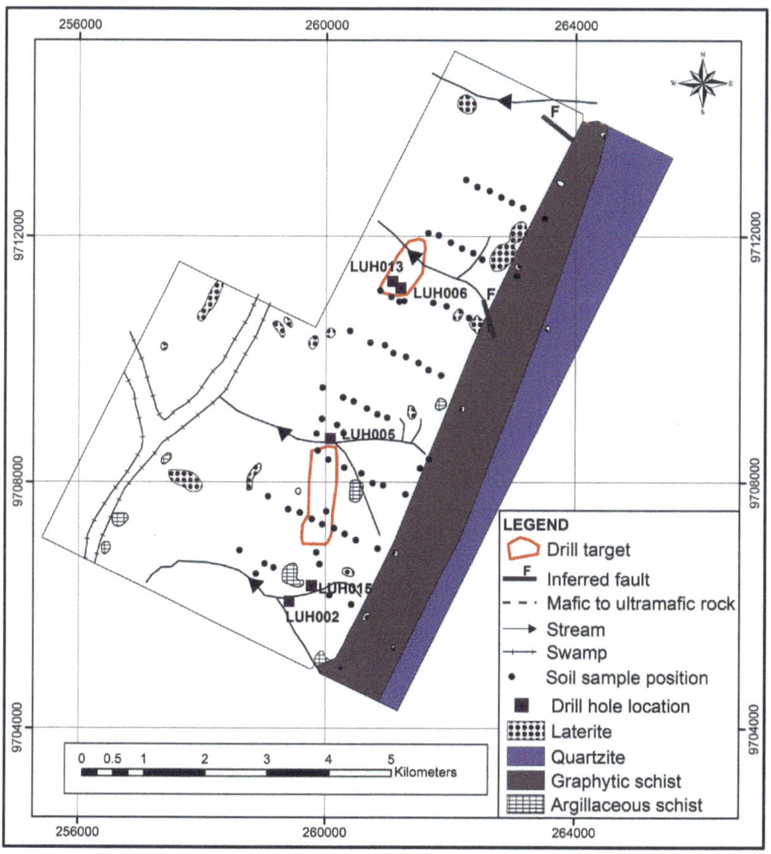

FIGURE 4.1.6 The overlay of the proposed drill targets on the geology of the Luhuma prospect. The relatively most potential area is within the area defined by coordinates UTM36 259799E/9707413N. Source: *From Macheyeki, A.S., 2011. Application of lithogeochemistry to exploration for Ni–Cu sulfide deposits in the Kabanga area, NW Tanzania.* J. Afr. Earth Sci. *61, 62-81.*

4.1.1.3 Limitations of vectors

Element ratios are useful if and only if they are obtained based on geochemical characteristics of elements within various minerals and not just the numbers. The emphasis here is for a scientist or researcher to understand well the element compositions of various minerals and to be able to classify elements; chalcophile, siderophiles, etc., so that whatever patterns are obtained in the process of element groupings can be easily interpreted.

Regarding the application of elements, it is important to establish the upper and lower limits of particular element ratios or vectors (e.g., Macheyeki, 2003). According to Macheyeki (2003), Fig. 4.1.5A and B, the upper and lower limit of $(Pd/V) \times 1000$ are 1000 and 195, respectively. However, the upper and lower limits alone are not enough to delineate ore potential areas but the patters of clustering of the vector values (Macheyeki, 2011). Further, one has to understand the geological setting in which a particular vector or set of vectors/element ratios are to be applied in order to bring about the expected results. The above vectors and their limits are suitable for the Mesoproterozoic Karagwe–Ankolean tectonic setting—they can be applied in similar geological settings.

References

Brand, N.W., 1999. Element ratios in nickel sulphide exploration: vectoring towards ore Environments. J. Geochem. Explor. 67, 145–165.

Chao, T.T., Sanzolone, R.F., 1992. Decomposition techniques. J. Geochem. Explor. 44, 65–106.

Evans, D.M., Byemelwa, L., Gilligan, J., 1999. Variability of magmatic sulphide compositions at the Kabanga nickel prospect, Tanzania. J. Afr. Earth Sci. 29 (2), 329–351. Available from: https://doi.org/10.1016/s0899-5362(99)00101-3.

Evans, D.M., Boad, I., Byemelwa, L., Gilligan, J., Kabete, J., Marcet, P., 2000. Kabanga magmatic nickel sulfide deposits, Tanzania: morphology and geochemistry of associated intrusions. J. Afr. Earth Sci. 30 (3), 651–674.

Hall, G.E.M., Bonham-Carter, G.F., 1988. Review of methods to determine gold, platinum and palladium in production-oriented geochemical laboratories, with application of statistical procedure to test bias. J. Geochem. Explor. 30, 255–286.

Ikingura, J.R., Reynolds, P.H., Watkinson, D.H., Bell, K., 1992. $^{40}Ar/^{39}Ar$ dating of granites of NE Kibaran Belt (Karagwe-Ankolean), northwestern Tanzania. J. Afr. Earth Sci. 15 (3/4), 501–511.

Macheyeki, A.S., 2003. Lithogeochemical vectors associated with the Kabanga Ni–Cu sulfide deposits, NW Tanzania. Unpublished M.Sc. Thesis. The University of the Western Cape, Cape Town, 117 p.

Macheyeki, A.S., 2011. Application of lithogeochemistry to exploration for Ni–Cu sulfide deposits in the Kabanga area, NW Tanzania. J. Afr. Earth Sci. 61, 62–81.

Tack, L., Liegeois, J.P., Deblond, A., Duchesne, J.C., 1994. Kibaran A-type granitoids and mafic rocks generated by two mantle sources in a late erogenic setting (Burundi). Precambrian Res. 68, 323–356.

Tissot, F., Swager, Ingovatov, A., van Straaten, H.P., Berg, R.C., Karbovsky, E., 1980. Mineralization in the Karagwe–Ankolean system of north-west Tanzania. Nat. Resour. Forum. 4, 85–94.

Further reading

Evans, D.M., 2017. Chromite compositions in nickel sulphide mineralized intrusions of the Kabanga-Musongati-Kapalagulu alignment, East Africa: petrologic and exploration significance. Ore Geol. Rev. 90, 307–321. Available from: https://doi.org/10.1016/j.oregeorev.2017.03.012.

SUBCHAPTER

4.2

In Asia

SUBCHAPTER

4.2.1

The use of mineral prospectivity modeling at orefield scale and prospect scale, Yangtze Metallogenic Belt, China

Xiaohui Li[1], *Feng Yuan*[1], *Mingming Zhang*[1], *Taofa Zhou*[1], *Simon M. Jowitt*[2] *and Alison Ord*[3]

[1]School of Resources and Environmental Engineering, Hefei University of Technology, Hefei, Anhui, P.R. China, [2]Department of Geoscience, University of Nevada–Las Vegas, Las Vegas, NV, United States, [3]Centre for Exploration Targeting, University of Western Australia, Crawley, WA, Australia

4.2.1.1 Regional geological setting

The middle–lower Yangtze Metallogenic Belt (MLYB) is one of the most important Fe–Cu–Au-polymetallic metallogenic belts in China and located within the Yangtze tectonic depression along the northern margin of the Yangtze Block, which formed between 149 and 105 Ma (Chang et al., 1991; Mao et al., 2011; Zhai et al., 1992; Zhou et al., 2015). It is bounded by the Xiangfan–Guangji and Tan–Lu faults to the northwest and by the Yangxin–Changzhou fault to the southeast (Fig. 4.2.1). Prolonged and related tectonic, magmatic, and metallogenic events since the Jinningian (c.850–800 Ma) have generated a faulted uplift and depression tectonic framework in this region as well as more than 200 diverse Fe–Cu–Au–polymetallic ore deposits and prospects that are located in a number of mining districts known as ore concentration areas. These include the volcanic rift basin-hosted Luzong and Ningwu OCAs and the secondary uplift-hosted Tongling, Anqing-Quichi, Jiurui, Ningzhen, and Edongnan OCAs. It is characterized by different kinds of magmatism in rift basins and uplift areas: uplift areas (e.g., Tongling) contain high-K calc-alkaline rock assemblage (including pyroxene

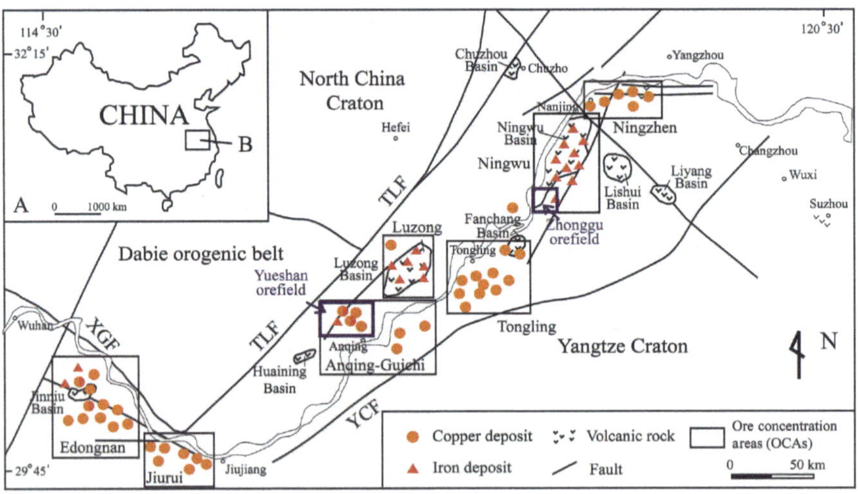

FIGURE 4.2.1 Map showing the (A) location of the middle and lower regions of the Yangtze Metallogenic Belt, and (B) the location of Yueshan orefield, volcanic basins, and ore concentration areas (OCAs) within the middle and lower Yangtze River Metallogenic Belt as well as the location of major settlements, faults, and major tectonic features. Source: *Modified after Chang, Y., Liu, X., Wu, Y., 1991. The Cu, Fe Metallogenic Belt in the Middle–Lower Reaches of Yangtze River. Geological Publish House, Beijing, 379 p. (in Chinese); Mao, J., Xie, G., Duan, C., Pirajno, F., Ishiyama, D., Chen, Y., 2011. A tectono-genetic model For Porphyry–Skarn–Stratabound Cu–Au–Mo–Fe and magnetite-apatite deposits along the middle-lower Yangtze River Valley, Eastern China. Ore Geol. Rev. 43, 294–314.*

monzodiorite, diorite, and granodiorite), whereas rift basins (e.g., Ningwu and Lucong) contain high-Na calc-alkaline plutonic rocks (e.g., pyroxene diorite and monzonite) and volcanic assemblage (e.g., trachyandesite, trachyte, andesite, and basalt). The MLYB contains diverse ore deposit types (e.g., Chang et al., 1991) that are dominated by the porphyry-, skarn-, and magnetite–apatite-type deposits that host the majority of the Cu–Fe–Au mineralization in this area (Fig. 4.2.1), all of which are closely associated with the extensive and voluminous Yanshanian tectono-thermal events recorded in this region (Sun et al., 2017; Yuan et al., 2012a; Zhou et al., 2015).

Here, a study based on multifractal analysis will be presented to show the usefulness of the multifractal analysis for the geochemical exploration data.

4.2.1.2 At orefield scale

1. Geology of the Yueshan orefield

   The Yueshan orefield is located in the northwest of the Anqing–Guichi OCA (Fig. 4.2.1) and records an intense history of folding, faulting, and magmatism (Fig. 4.2.2). The area contains Silurian to Jurassic sedimentary rocks (Fig. 4.2.2), and the mineralization and magmatism in this area were controlled by NNW–SSE trending faults as well as E–W trending faults within Precambrian Dongling Formation (1895 ± 38 Ma) basement in this area (Xing et al., 1994; Zhou et al., 2007). The majority of the widespread magmatic rocks in this area were emplaced during the Yanshanian, including the multiphase diorite–quartz diorite–monzodiorite Yueshan, Zongpu, and Wuheng intrusions (Fig. 4.2.2). The Yueshan orefield contains numerous skarn-type Fe–Cu deposits, the majority of which are associated with the Yueshan intrusion (Fig. 4.2.2). The mineralization associated with these deposits is concentrated within the contact zone between the Yueshan intrusion and the surrounding Triassic sedimentary carbonates of the Yueshan (T_2y) and Nanlinhu (T_1n) formations. The two largest deposits within the orefield are both skarn-type deposits, namely the Anqing Fe–Cu deposit and the Zhuchong Fe–Cu deposit (326 Geological Team, 2012).
2. Exploration model

   Because skarn-type Fe–Cu deposits dominate the Yueshan orefield, the exploration model represents a conceptual model that embodies the descriptive features of skarn-type mineralization as

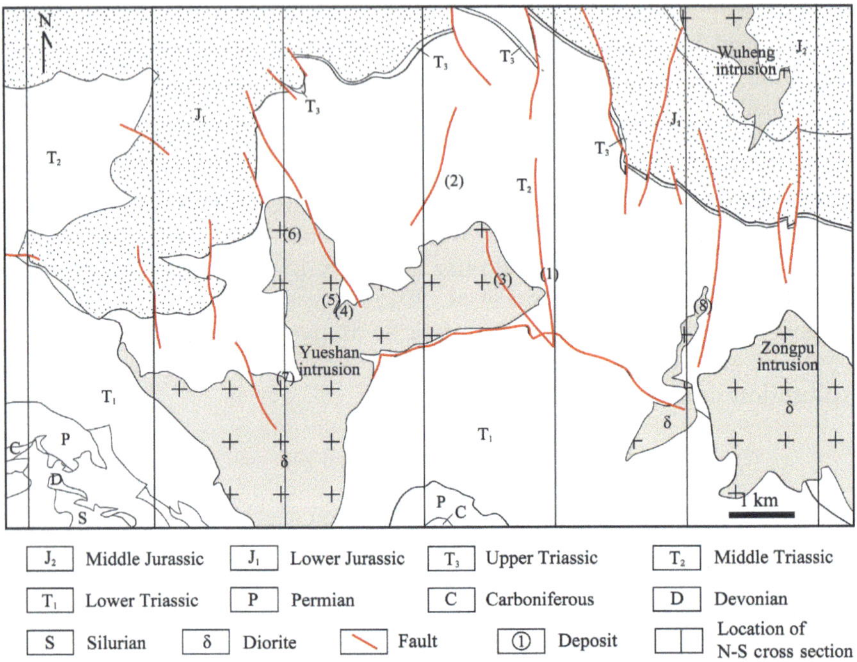

FIGURE 4.2.2 Geological map of the Yueshan orefield, showing the main geological units, major faults, and the location of important mineral deposits as follows: (1) Anqing Cu–Fe deposit, (2) Zhuchong Fe–Cu deposit, (3) Matoushan Cu–Fe deposit, (4) Longmenshan Cu deposit, (5) Liujiawa Fe–Cu deposit, (6) Tongniujin Cu–Mo deposit, (7) Liujiadapai Fe deposit, (8) Bailin Au deposit.

well as enabling the definition of exploration criteria for prospectivity modeling. The most important exploration criteria used within our exploration model are as follows:

a. Sedimentary units

The skarn-type mineralization within the Yueshan orefield is associated with dolomite and limestone units that preferentially host mineralization in the study area. These middle Carboniferous to Middle Triassic rocks are littoral to neritic carbonates that are interbedded with bathyal facies sediments and alternate with marine to continental clastic sediments (Liu et al., 2011).

b. Intrusions

Numerous ~140 Ma diorite–quartz diorite–monzodiorite intrusions are located within the Yueshan orefield, including the Yueshan, Zongpu, and Wuheng intrusions. The skarn-type deposits within the Yueshan orefield are closely associated with dioritic composition intrusions.

c. Structures

The Fe–Cu mineralization is preferentially located close to faults and is associated with domes, depressions, and roof segments within the diorite intrusions as well as tongue-shaped contact zones between Triassic sediments and diorite intrusions, all of which are used as exploration criteria in this study.

Previous research on computational simulations indicates that the orebodies within the Anqing deposit are mostly localized in high dilation zones as a result of incremental volumetric strain (Liu et al., 2011, 2012). The mechano-thermo-hydrological (MTH) processes that operate in these areas cause rocks to deform, dilate, and eventually fail. This in turn can induce incremental increases in volumetric strain that can facilitate the deposition of ores or the formation of space conducive to ore deposition (Liu et al., 2012). This indicates the importance of incorporating these regions in prospectivity modeling.

d. Geophysical characteristics

The carbonates in the study area have higher average densities than the diorite intrusions ($\sim 2.67 \times 10^3$ kg/m^3) in this region as well as the sandstone and siltstone units in this area. This means that gravity geophysical data can be used to identify the majority of the mineralization-related rocks in the study area. In addition, the dioritic intrusions have average magnetic susceptibility values of $\sim 4000 \times 10^{-5}$ SI, whereas the magnetite mineralization in this area has an average magnetic susceptibility value of around $80{,}000 \times 10^{-5}$ SI. This means that both diorite intrusions and shallow magnetite mineralization can be magnetically detected and potentially discriminated between in the study area. However, the depth of the intrusions and mineralization also plays a role in that deep-seated diorite intrusions may be associated with weakly magnetic anomalies at the surface, and deep-seated magnetite mineralization is difficult to detect in this region as a result of the magnetic susceptibility of the diorite intrusions. Therefore, deep-seated Zhuchong Fe–Cu deposit has not been discovered until 2012 by the use of controlled-source audio-frequency magnetotelluric data to identify geological structures and low resistivity anomalies at depth in the study area.

3. 3D geological modeling

The 3D mineral prospectivity modeling is based on 3D geological models that form the basis of all of the 3D predictive maps generated during this. This approach to the 3D geological modeling was outlined in detail by Malehmir et al. (2009) and Lü et al. (2012) and involves five steps, namely (1) defining the 3D

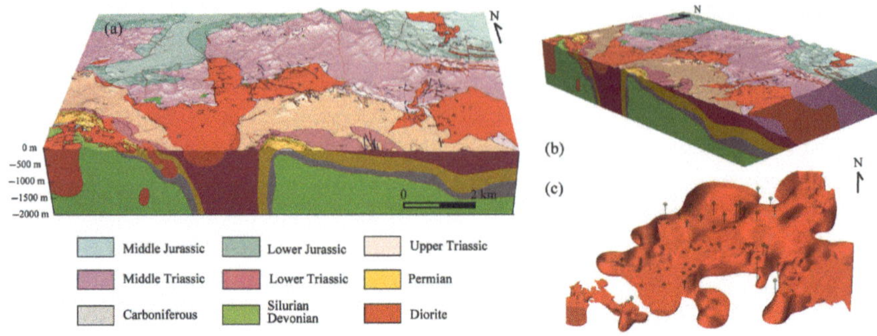

FIGURE 4.2.3 3D geological model of the Yueshan orefield generated during this study, including (A) front view of the 3D geological model, (B) oblique view of the 3D geological model, and (C) 3D model for the diorite intrusion in this area incorporating drill hole data as further constraints.

limits of the model area, (2) constructing a series of 2D cross-sections over the model area, (3) integration and interpretation of these 2D geological cross-sections with surface geology and borehole data, (4) geophysical forward modeling of the 2D cross-sections, and (5) final data integration and 3D visualization.

An implicit 3D geological modeling approach (Calcagno et al., 2008; Jessell, 2001) that employs Geomodelle software was used to incorporate data from geological maps and cross-sections with the geophysical models and drill hole data (Fig. 4.2.3).

4. D predictive map generation

a. 3D computational simulations

Computational simulation methods have been widely used during investigations into the processes involved in the generation of complex magmato-hydrothermal (e.g., skarn-type) mineral deposits, and recent advances in this area include the 3D modeling of these systems (Liu et al., 2012; Murphy et al., 2008; Ord et al., 2012). Computational simulation methods can provide insights into the dynamic and genetic relationships between igneous intrusions and skarn-type mineralization (as well as other classes of mineral deposit) using approaches such as coupled simulating deformation and the modeling of heat transfer and fluid flow during mineralization and the cooling of ore-related intrusions (MTH).

In this study, computational simulation methods were used to simulate MTH processes during the syn-extensional cooling of the intrusion in order to identify concealed high dilation zones that are known to preferentially host magmato-hydrothermal mineralization (Liu et al., 2011). The intrusion and the surrounding wall rocks were defined as

Mohr–Coulomb materials where the governing equations used for simulating MTH processes are described as follows (FLAC3D User's Guide, 2012):

$$q^W = -k\nabla(P - \rho_w g \cdot x) \tag{4.2.1}$$

Eq. (4.2.1) is Darcy's law describing fluid flow, where q^W is fluid specific discharge, k is the fluid mobility coefficient, P is the pressure of the pore fluid, ρ_w is fluid density, and g is gravitational acceleration.

$$q^T = -k^T \nabla T \tag{4.2.2}$$

Eq. (4.2.2) is Fourier's law describing heat conduction, where q^T is heat flux, k^T is the effective thermal conductivity, and T is temperature.

$$\frac{\partial \zeta}{\partial t} = -q_{i,i} + q_v^W \tag{4.2.3}$$

Eq. (4.2.3) describes the conservation of mass, where ζ is the variation of fluid content, q_i is the fluid specific-discharge vector in the x_i direction, and q_v^W is the volumetric fluid source.

$$C^T \frac{\partial T}{\partial t} + \nabla q^T + \rho_0 c_w q^W \cdot \nabla \mathrm{T} - q_v^T = 0 \tag{4.2.4}$$

Eq. (4.2.4) describes the energy balance for convective–diffusive heat transport, where C^T is the effective specific heat, T is temperature, q^T is heat flux, ρ_0 and C_w are the reference density and the specific heat of the fluid, respectively, and q_v^T is the volumetric fluid source.

$$\rho \frac{dv_i}{dt} = \sigma_{ij,j} + \rho g_i \tag{4.2.5}$$

Eq. (4.2.5) describes the conservation of momentum, where v_i is the velocity component in the x_i direction, ρ is the bulk density of the porous medium, σ_{ij} is the stress tensor of the solid, and g_i is the component of gravitational acceleration in the x_i direction.

$$\frac{\partial \varepsilon_{ij}}{\partial t} = \alpha_t \frac{\partial T}{\partial t} \delta_{ij} \tag{4.2.6}$$

$$\frac{\partial P}{\partial t} = M\left(\frac{\partial \zeta}{\partial t} - \alpha \frac{\partial \varepsilon}{\partial t} + \beta \frac{\partial T}{\partial t}\right) \tag{4.2.7}$$

Finally, Eqs. (4.2.6) and (4.2.7) describe the coupled MTH constitutive relation, where ε_{ij} is the thermal strain tensor, T is temperature, α_t is the coefficient of linear thermal expansion, δ_{ij} is the Kronecker delta, ζ is the variation in fluid content, M is the Biot modulus, α is the Biot coefficient, ε is the volumetric strain, and β is the volumetric thermal expansion of the porous matrix.

This study uses FLAC3D, a 3D finite-difference software, to solve the governing equations used for simulating MTH processes within discrete 3D geological models (Itasca Consulting Group, 2012). The 3D computational simulations undertaken during this study were split into three steps as follows:

a. Construction of a 3D geological model for 3D computational simulations

The fact that implicit 3D geological modeling is based on discrete 3D cells means that this approach can directly provide discrete 3D geological models for use with the FLAC3D software. Meanwhile, previous research indicates that the ore-forming fluids that generated the skarn-type Anqing deposit (currently located at depths of -280 to -580 m) were under pressures of around 8×10^7 Pa (Zhou et al., 2007). Assuming these ore-forming fluids were under lithostatic pressure indicates that these ore bodies were originally located some -3000 m below the surface. As such, the geological model was reconstructed for 3D computational simulations by adding some 2500 m of Jurassic sedimentary units to the top of the model (Fig. 4.2.4).

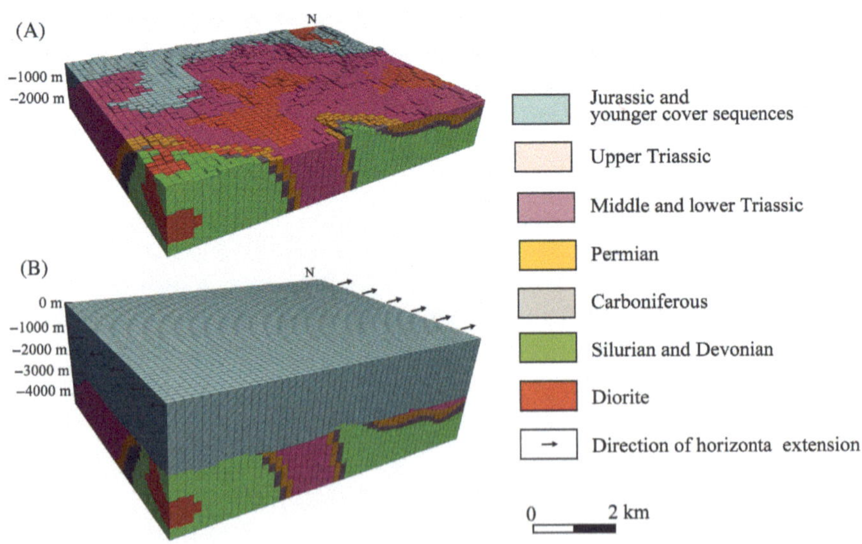

FIGURE 4.2.4 3D geological models used during 3D computational simulations (A) without 2500 m Jurassic cover and (B) covered by 2500 m of Jurassic sediments.

b. The setting of properties and boundary conditions

The initial temperature of the top surface is set at 25°C and is kept constant with an initial temperature gradient of 20°C/km for the sedimentary section and an initial intrusion temperature of 700°C. The initial pore pressure of the sedimentary rocks is set to hydrostatic conditions, whereas the initial pore pressure of the intrusion is set to be near lithostatic.

The timing of formation of the diorite intrusion and associated mineralization within the Yueshan orefield was associated with a change from compressional to an extensional tectonic regime in the lower Yangtze River region (Dong and Qiu, 1993; Mao et al., 2006). As such, our models are set to horizontally extend with a symmetrical boundary velocity of 2.4×10^{-10} ms^{-1} at both sides parallel to the E−W direction of the model to represent this tectonic extension (Liu et al., 2012; Fig. 4.2.4).

c. 3D computational simulation

Previous research indicates that the ore bodies within the Anqing deposit are mostly located in high dilation zones as a result of incremental volumetric strain, a process that can facilitate the deposition of ores or the formation of space conducive to ore formation (Liu et al., 2012). This means that incremental increases in volumetric strain represent an exploration criterion that can be used in 3D prospectivity modeling.

The simulation ran for a thousand years and generated the major ore bodies within the Anqing and Zhuchong deposit, all of which are located at different depths within high dilation zones (i.e., high incremental volumetric strain increase zones) along the contact zone between the intrusion and the surrounding Triassic country rocks (Fig. 4.2.5). Simulation result not only clearly delineates the relationship between high dilation zones and known mineral deposits but also highlights some hitherto unknown high dilation zones within the margin of the intrusion (Fig. 4.2.6A). This means that this simulation-based approach could be a useful tool in terms of identifying areas for future exploration. In addition, this simulation highlights numerous high dilation zones in the study area, meaning that the results of this simulation results can be combined with other predictive maps to reduce exploration uncertainty in this region.

3D distance fields (or 3D buffers) can not only allow the identification of spatial relationships between mineralization and individual mineral exploration criteria as well as providing 3D predictive maps for 3D data fusion and

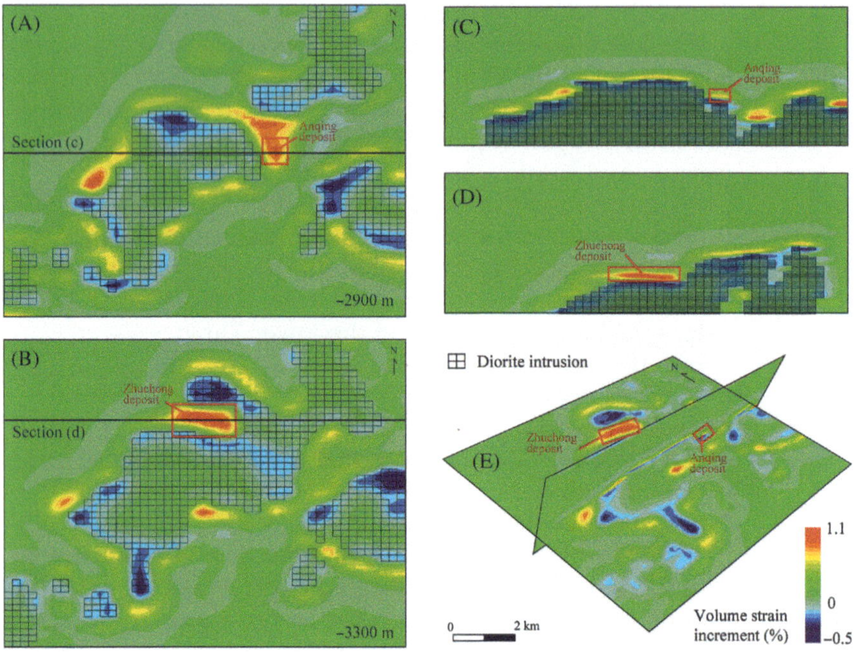

FIGURE 4.2.5 Changes in incremental volume strain after 3D computational simulation shown on (A) plan view at a depth of −2900 m (equal to −400 m at the present day), (B) plan view at −3300 m (equal to −800 m at the present day), (C) cross-section across the Anqing deposit, (D) cross-section across the Zhuchong deposit, and (E) a 3D overlay of cross-section and plan views.

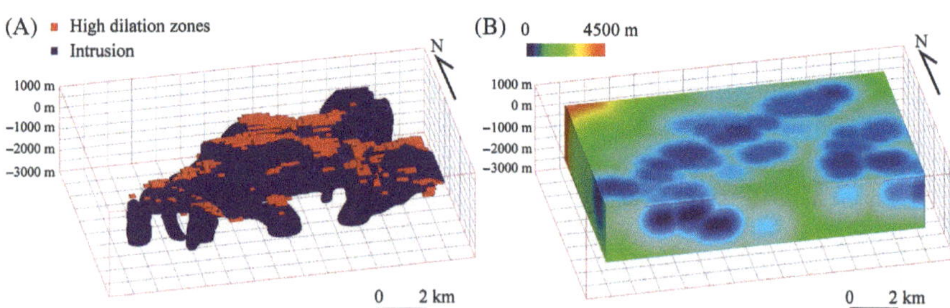

FIGURE 4.2.6 (A) Location of blocks within high dilation zones in the vicinity of the margins of the diorite intrusion and (B) 3D model showing variations in spatial proximity to high dilation zones.

targeting, but also can decrease the uncertainties caused by inaccuracies within 3D geological models, theoretical limitations and the precision of 3D computational simulations, 3D spatial analysis, and 3D physical property inversions. Here, a 3D signed Euclidean distance transformation method that can enable significant increases in computational efficiency (Lin and Wang, 2003; Jones et al., 2006; Li et al., 2016) was used to initially calculate 3D distance fields for the location of high dilation zones (Fig. 4.2.6B). The 3D distance fields for the remaining exploration criteria were then generated using 3D spatial analysis, with later inversion of the 3D physical properties in the study area.

b. 3D spatial analysis

In this study, 3D spatial analysis methods including 3D mathematical morphology and Gaussian curvature calculation method were used to define and extract the mineral exploration criteria described in Section 4.2.1.1 (and based on the exploration model outlined earlier) from the 3D geological model constructed during this study.

The sedimentary units, intrusions, faults, doming, and depressions within the contact between the diorite intrusion and the surrounding country rock and tongue-shaped sedimentary units which were related to skarn-type mineralization were defined and extracted (Figs. 4.2.7–4.2.11).

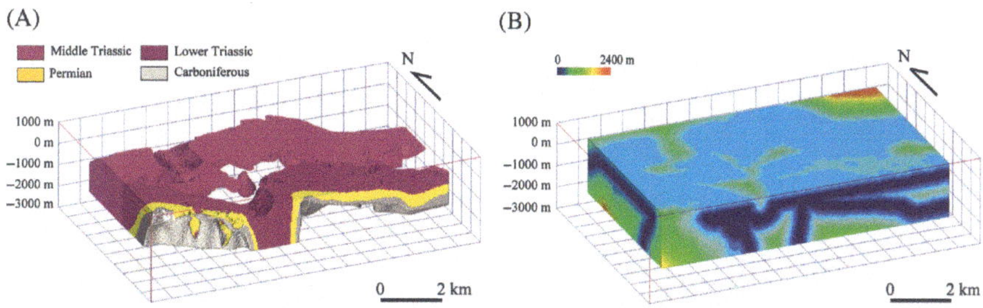

FIGURE 4.2.7 (A) 3D model of the Carboniferous to Middle Triassic sedimentary units in the study area and (B) 3D model showing variations in proximity to the contact surface of these sedimentary units.

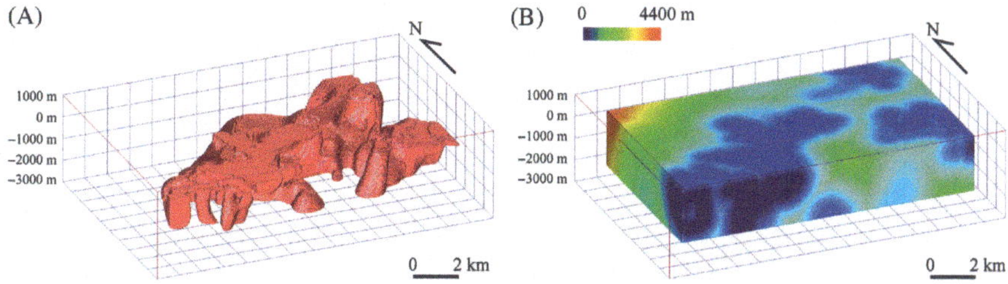

FIGURE 4.2.8 (A) 3D model of the diorite intrusion and (B) 3D model showing variations in proximity to the contact surface of diorite intrusion within the study area.

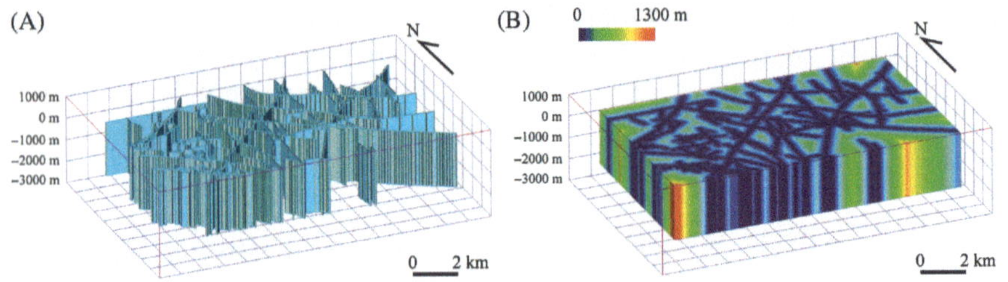

FIGURE 4.2.9 (A) 3D model of faults within the study area faults and (B) 3D model showing variations in the proximity to faults within the study area.

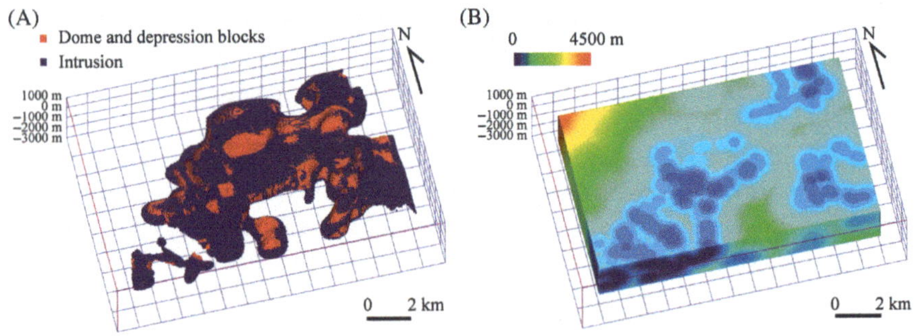

FIGURE 4.2.10 (A) Location of blocks within domes and depressions in the diorite intrusion contact surface and (B) 3D model showing variations in spatial proximity to dome and depression zones within the diorite intrusion contact.

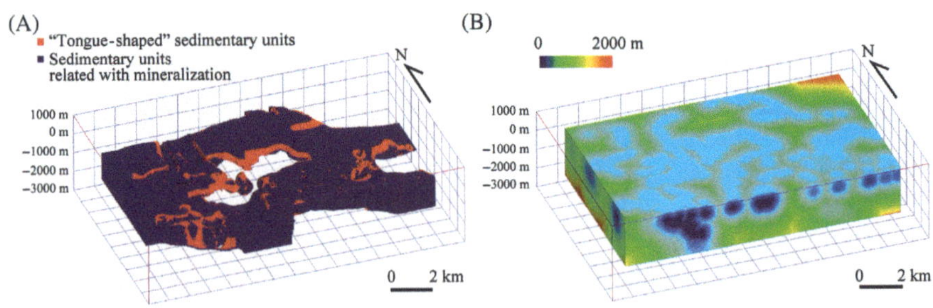

FIGURE 4.2.11 (A) Location of blocks within tongue-shaped sedimentary units and (B) 3D model showing variations in proximity to these blocks within tongue-shaped sedimentary units.

c. 3D Physical property inversion

Li and Oldenburg (1996, 1998) outlined a geologically constrained 3D property inversion algorithm, which uses a model objective function that includes a term that measures the difference between the recovered model and a reference model, and how the difference between recovered and reference models varies between cell in each of three orthogonal directions (Williams, 2008). The model objective function is as follows:

$$\begin{aligned}\phi_m m = {} & \alpha_s \int_V w_s [w_r(z)(m - m_{\text{ref}})]^2 dV + \\ & \alpha_x \int_V w_x \left[\frac{\partial (w_r(z)(m - m_{\text{ref}}))}{\partial x}\right]^2 dV + \\ & \alpha_y \int_V w_y \left[\frac{\partial (w_r(z)(m - m_{\text{ref}}))}{\partial y}\right]^2 dV + \\ & \alpha_z \int_V w_z \left[\frac{\partial (w_r(z)(m - m_{\text{ref}}))}{\partial z}\right]^2 dV +\end{aligned} \tag{4.2.8}$$

where m is the recovered model and m_{ref} is the reference model. The first integral component measures the smallness and last three integral components measure the smoothness of the difference between m and m_{ref}, α_s, α_x, α_y, and α_z are used to balance the contributions of the smallness and smoothness components, w_s, w_x, w_y, and w_z are weighting functions and are spatially dependent, and the function $w_r(z)$ is a depth weighting to counteract the decay of the potential-field response with distance from the source (Williams, 2008).

The geologically constrained 3D property inversion algorithm mentioned earlier was used to inversely calculate the density and susceptibility of individual cells within the study area. The gravity data used for inversion in this paper was measured on a 200×200 m grid, with magnetic data measured on a 100×100 m grid and a reference model built using geological mapping and physical property information. In the study area, high density (>2.67 g/cm^{3}density) cells most likely represent mineralization-related carbonate, dolomite, limestone, marble, or magnetite units and high magnetic susceptibility ($>4000 \times 10^{-5}$ SI) cells most likely correspond to the diorite intrusion and magnetite. Therefore, this means that the high density and magnetic susceptibility cells related to Fe–Cu mineralization can be further extracted in 3D using 3D geophysical inversion models, respectively (Fig. 4.2.12).

5. 3D data fusion and mineral prospectivity mapping

Data-driven models have a number of advantages over knowledge-driven models during Brownfield exploration, the most important of which is the fact that data-driven models can objectively identify and quantify spatial relationships between data inputs and mineralization.

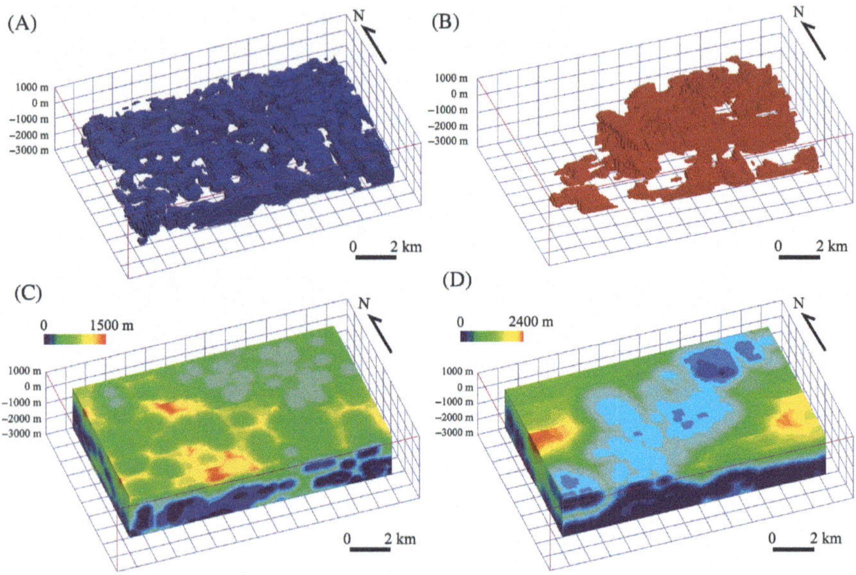

FIGURE 4.2.12 3D models showing the location of (A) high density and (B) high magnetic susceptibility geological bodies and variations in proximity to blocks within (C) high density and (D) high magnetic susceptibility geological body contact surfaces.

In this study, data-driven logistic regression (Chung and Agterberg, 1980) was used to integrate all of the 3D predictive maps and highlight exploration targets at depth; both of these approaches have recently been used in 3D prospectivity modeling (Payne et al., 2015; Yuan et al., 2014; Li et al., 2015, 2019). Logistic regression method can be expressed as follows:

$$\pi(d) = \frac{e^{\alpha+\beta_i x_i}}{1 + e^{\alpha+\beta_i x_i}}$$

where $\pi(d)$ is the probability of a mineralized occurrence, x_i is the ith ($i = 1$ to N) predictive map, α is a constant, and β_i is the regression coefficient for x_i.

The results of 3D prospectivity modeling are shown in Fig. 4.2.13. The results show that high-prospectivity areas not only delineated areas of known mineralization but also identified several new exploration targets (1–7 in Fig. 4.2.13) outside of any areas containing known mineralization.

These include new targets at an average depth of 1300 m beneath the known Bailin Au deposit and associated with the contact between Carboniferous sedimentary formations and diorite intrusions, all of which highlight the fact that the Yueshan orefield remains highly prospective for exploration for hitherto undiscovered deep-seated skarn-type mineralization.

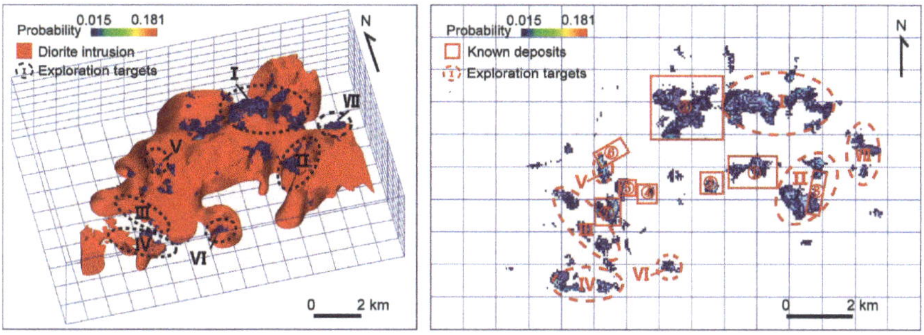

FIGURE 4.2.13 High-prospectivity areas and exploration targets for skarn-type Fe–Cu mineralization identified during this study. Mineral deposit locations are as follows: (1) Anqing Cu–Fe deposit, (2) Zhuchong Fe–Cu deposit, (3) Matoushan Cu–Fe deposit, (4) Longmenshan Cu deposit, (5) Liujiawa Fe–Cu deposit, (6) Tongniujin Cu–Mo deposit, (7) Liujiadapai Fe deposit, and (8) Bailin Au deposit.

4.2.1.3 At prospect scale

The methodology and workflow of 3D mineral prospectivity modeling can be used not only at orefield scale but also at prospect scale. Here, a case study on Baixiangshan deposit will be introduced.

1. Geological setting
   a. Geology of the Zhonggu orefield

      The Zhonggu orefield is located in the south of the Ningwu volcanic basin, eastern MLYB (Fig. 4.2.1) and has a history of intense folding and faulting and significant magmatism. Mineralization and magmatism in the area were controlled by basement-hosted NNE–SSW and WNW–ESE-trending fault sets. The sedimentary basement sequence within the orefield is subdivided into the Middle Triassic Zhouchongcun (T_2z) and Huangmaqing formations (T_3h) and the Lower–Middle Jurassic Xiangshan Group ($J_{1\text{-}2}\,xn$). Iron mineralization in this area is hosted by dolomitic limestones of the Middle Triassic Zhouchongcun Formation (T_2z) and sandstones and shales of the Huangmaqing (T_3h) Formation, with small hypabyssal mineralization-related subvolcanic intrusions prevalent in the area.
   b. Geology of the Baixiangshan Deposit

      The Baixiangshan iron deposit is associated with the dioritic Baixiangshan intrusion (Fig. 4.2.14A). The magmas that formed this intrusion were emplaced into a sedimentary sequence that, from top to bottom, consists of Cretaceous volcanic rocks, the Lower–Middle Jurassic Xiangshan Group, and the Upper Triassic

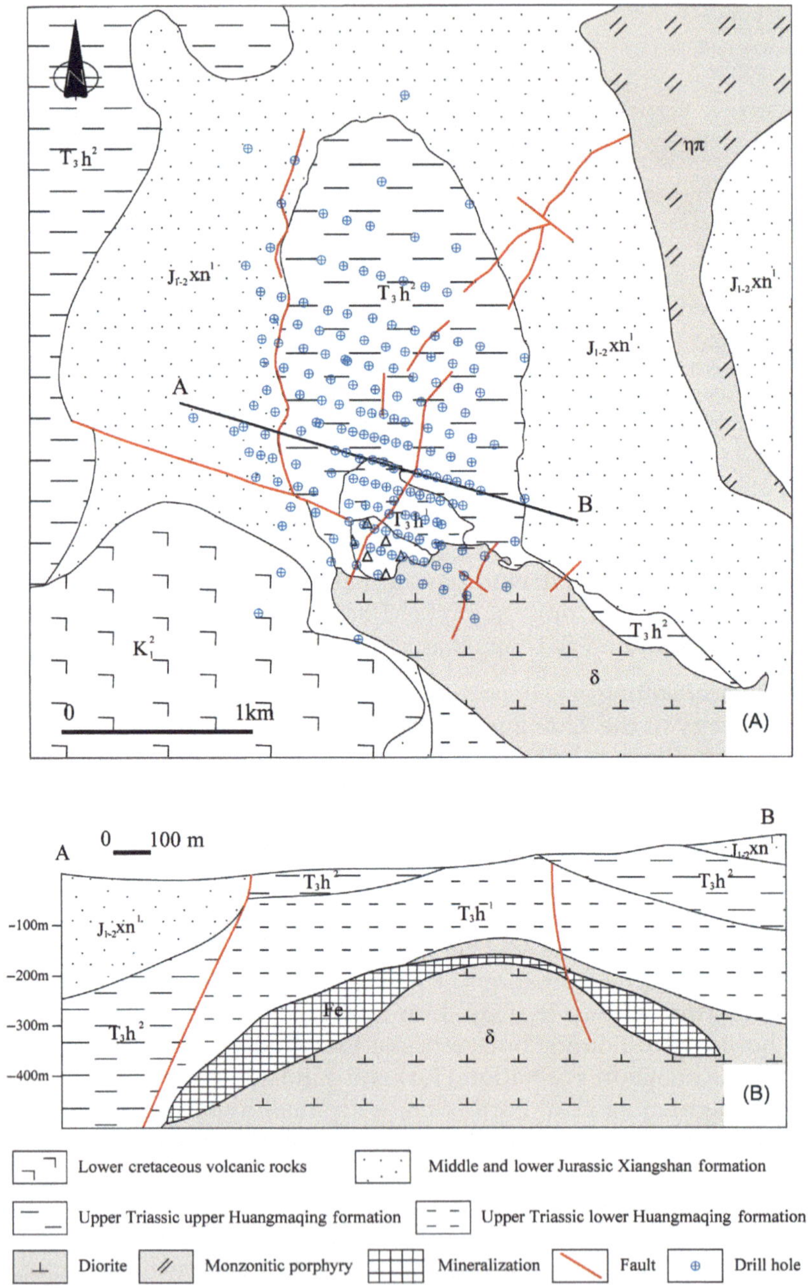

FIGURE 4.2.14 Map (A) and cross-section (B) showing the geology of the Baixiangshan deposit. Source: *Adapted from East China Metallurgical Bureau of Geology and Exploration, 1982. Geological report on the Baixiangshan deposit.*

Huangmaqing Formation, with emplacement of the diorite during the Early Cretaceous (~130 Ma; Zhou et al., 2011). The majority of the individual orebodies within the Baixiangshan deposit are located at the contact between the dioritic intrusion and the Triassic Lower Huangmaqing Formation, with mineralization apparently controlled by the contact between the two lithologies relative to the Baixiangshan anticline (Fig. 4.2.14B).

Ore minerals within the Baixiangshan deposit are dominated by magnetite with pseudomorphic and specular hematite after primary magnetite, and minor limonite and pyrite, within a gangue that includes apatite, diopside, quartz, kaolinite, albite, and actinolite. Individual ore bodies are often disseminated or banded, with massive and brecciated sections present in some areas (Fan et al., 2011).

2. Exploration model

The Baixiangshan deposit is associated with a dioritic intrusion, with the majority of mineralization occurring at the contact between the intrusion and the Triassic Lower Huangmaqing Formation. Given this, the bottom contact of the Lower Huangmaqing Formation, and the upper contact of the dioritic intrusion were defined as exploration criteria to be used during prospectivity modeling. In addition, as faults in the study area formed preferential pathways for both magmatism and, more importantly, hydrothermal fluids, so the faults were also defined as another exploration criterion.

The deposit and 3D geological model described earlier suggest that ore deposition generally occurred near areas that were uplifted during intrusion of the dioritic magma (i.e., within the roof zone of the intrusion, usually the locus of hydrothermal fluid flow and activity during intrusion), meaning that uplifted regions may also be used as an exploration criterion.

Magnetic surveying is an efficient and effective geophysical method for identifying the location of iron mineralization. Two-dimensional magnetic data for the study area indicate that the residual magnetic anomaly maximum of this area, a maximum value that is spatially associated with iron mineralization, is close to 1600 nT; this, combined with the near vertical nature of the intrusion, meant that 2D high residual magnetic anomalies were also used as an exploration criterion.

3. 3D geological modeling

Based on geological mapping, cross-section construction, and drill hole intercepts, a 3D geological model was developed for this study. Both this modeling and the definition of individual 3D geological bodies were undertaken using the Surpac software

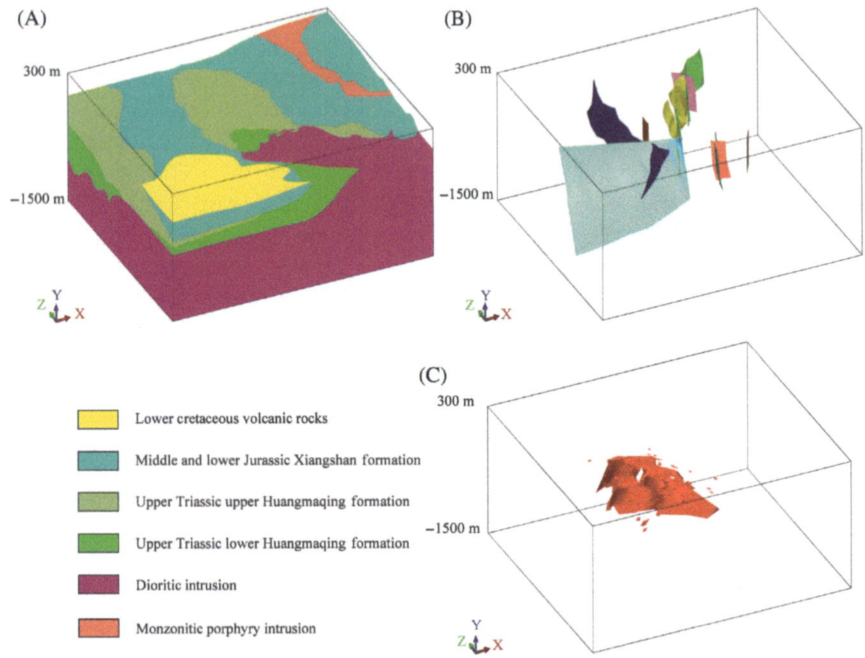

FIGURE 4.2.15 Diagrams showing the 3D geological models used during this study: (A) 3D stratigraphic model, (B) 3D fault model, where F1–F10 are faults within the Zhonggu basin, (C) 3D ore body model defined by a total Fe >30% ore shell.

package, and the 3D stratigraphic model, 3D fault model, and 3D ore body model developed for this study are shown in Fig. 4.2.15.

4. 3D predictive map generation

Predictive maps were generated using 3D block models and the 3D geological model described earlier to assign property values to individual blocks prior to spatial analysis.

As discussed earlier, the faults, bottom contact of the Lower Huangmaqing Formation, the upper contact of the dioritic intrusion, and the uplifted regions of the intrusion as exploration criteria were defined first. Then, 3D distance fields for each exploration criterion were constructed in order to determine the spatial range of the exploration criteria, yielding a series of predictive maps (Fig. 4.2.16).

5. 3D data fusion and prospectivity mapping

a. Weights of evidence

Weights-of-evidence modeling is a bivariate statistical approach based on Bayes' theory of conditional probability that quantifies the relationship between mapped distributions of a given dataset and training points, which in this case are areas

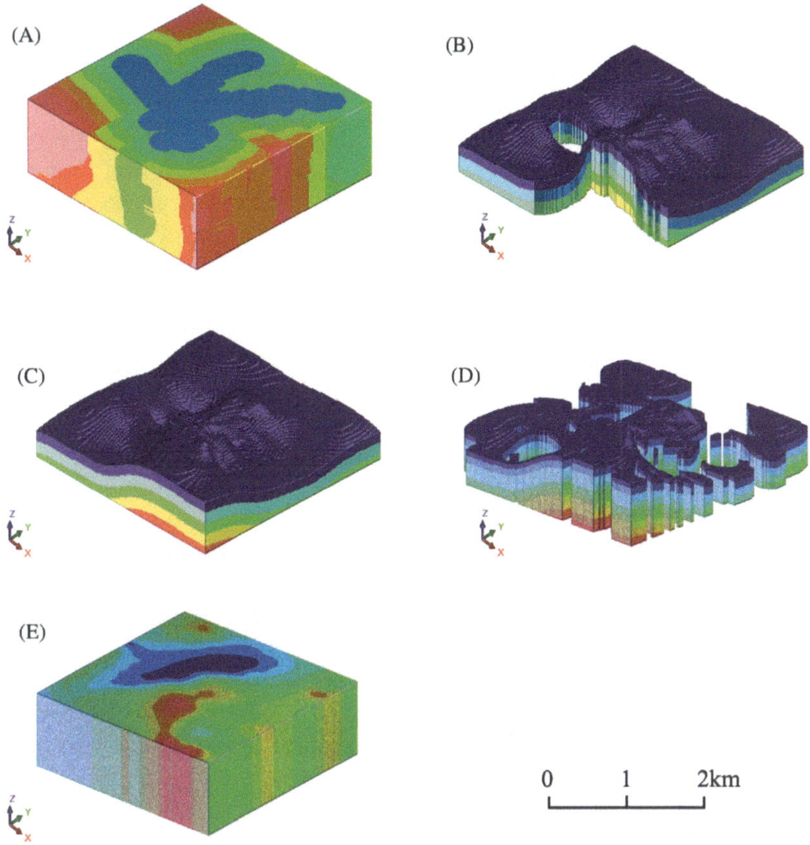

FIGURE 4.2.16 3D predictive maps:3D distance fields of (A) faults, (B) bottom contact of Lower Huangmaqing Formation, (C) top contact of the dioritic intrusion, and (D) uplifted blocks associated with intrusion of the dioritic magma, and (E) interpolation of unbuffered residual magnetic intensity data, showing the assignment of uniform values with depth due to the 2D nature of the magnetic data.

of known mineralization. This approach is widely used in 2D prospectivity modeling and combines exploration criteria with available data to derive posterior probability values for a given part of the model. These values represent the probability of mineralization occurring within a given block, based on the available data, training points (i.e., known mineralization) and the mineral deposit model being used. More details on the derivation of these formulae and the use of weights-of-evidence analysis are given in Bonham-Carter (1994).

b. Calculation of binary predictive maps

One of the most important steps in weights-of-evidence-based prospectivity modeling is the conversion of continuous data into discrete classes by the creation of binary maps. In this study, contrast values were calculated for each exploration criteria category, and all of the contrast curves defined for each individual exploration criteria have similar shapes and have rapid decreases in C values at a distance where $C=1$. Therefore categorical intervals of $C>1$ were classified as prospective areas within the optimum distance for binary map calculation and with blocks outside a distance defined by values of $C<1$ classified as unprospective.

c. Calculation of binary predictive maps

Exploration criteria-based binary predictive maps of exploration criteria were compared with training data to calculate positive (W^+) and negative (W^-) weights (Table 4.2.1). All exploration criteria barring residual magnetic intensity have positive (W^+) weights that are significantly larger than their negative (W^-) weights and have contrast values (C) larger than 2.8, indicating a clear positive spatial relationship between the exploration criteria used here and areas containing known mineralization.

The binary maps listed in Table 4.2.1 were used to create a posterior probability map, and exploration targets were demarcated using capture–efficiency curves (Fabbri and Chang-Jo, 2008; Porwal et al., 2010) to determine threshold

TABLE 4.2.1 Weight and contrast values determined during weights-of-evidence modeling.

Exploration criteria	Method of classification	Prospective range	Positive weighting	Negative weighting	Contrast
Faults	Buffering	0–180 (m)	1.524	− 1.29	2.814
Bottom contact of the Lower Huangmaqing Formation		0–80 (m)	2.545	− 1.64	4.185
Top contact of dioritic intrusion		0–80 (m)	1.775	− 1.419	3.195
Uplifted blocks associated with the dioritic intrusion		0–120 (m)	1.919	− 1.766	3.684
Residual magnetic intensity	Thresholding of classes	1640–980 (nT)	1.886	− 1.071	2.957

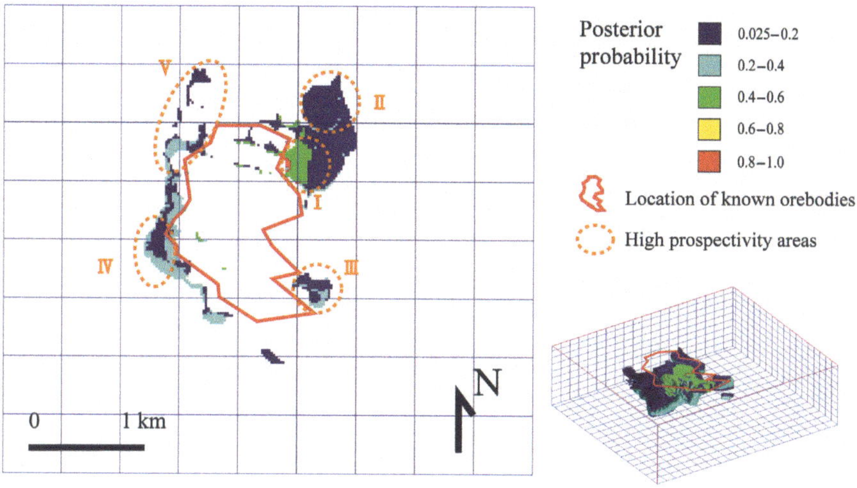

FIGURE 4.2.17 Distribution of high-prospectivity areas defined by the modeling undertaken during this study.

values that effectively separate high- and low-prospectivity areas. The 3D prospectivity modeling undertaken during this study successfully predicts the majority of the known mineralization in the study area and suggests that the approaches used here are applicable to Baixiangshan mining area (Fig. 4.2.17).

Acknowledgments

The researches introduced in this chapter come from two published papers as follows:

Li, X., Yuan, F., Zhang, M., Jowitt, S.M., Ord, A., Zhou, T., Dai, W., 2019. 3D computational simulation-based mineral prospectivity modeling for exploration for concealed Fe–Cu skarn-type mineralization within the Yueshan ore field, Anqing district, Anhui Province, China. Ore Geol. Rev. 105, 1–17.

Yuan, F., Li, X., Zhang, M., Jowitt, S.M., Jia, C., Zheng, T., Zhou, T., 2014. Three-dimensional weights of evidence-based prospectivity modeling: a case study of the Baixiangshan mining area, Ningwu Basin, Middle and Lower Yangtze Metallogenic Belt, China. J. Geochem. Explor. 145, 82–97.

The researches were financially supported by funds from the National Natural Science Foundation of China (Grant Nos. 41820104007, 41702353, 41672069, 41872247), the National Key R&D Program of China (Grant Nos. 2016YFC0600209, 2016YFC0600206), the Public Geological Survey Program of Anhui Province (Grant No. 2009–13), and the Oversea Master Program (Grant No. MS2015HFGY026).

SUBCHAPTER

4.3.1

Multifractal modeling of geochemical anomalies associated with Cu and Au mineralization in the NW Junggar area, northern Xinjiang Province, China

Feng Yuan[1], *Xiaohui Li*[1], *Mingming Zhang*[1], *Taofa Zhou*[1] *and Simon M. Jowitt*[2]

[1]School of Resources and Environmental Engineering, Hefei University of Technology, Hefei, Anhui, P.R. China, [2]Department of Geoscience, University of Nevada–Las Vegas, Las Vegas, NV, United States

Exploration geochemistry is an important method in mineral exploration targeting and the determination of the prospectivity of an area during mineral exploration. This is especially true of areas that are relatively unexplored, as potential mineralization-related geochemical anomalies can be rapidly delineated, meaning that exploration geochemistry data can be used as a guide for targeting areas for more detailed exploration (Bonham-Carter, 1994; Pan and Harris, 2000). In addition, these delineated anomalies can efficiently be used in more advanced exploration-focused modeling, such as GIS-based prospectivity mapping (Carranza, 2009; Wang et al., 2011).

Recent research has identified that the release of energy or the accumulation of mass during various geological processes, such as volcanic eruptions, occurs within narrow intervals in time or space, leading to the formation of geological anomalies (Cheng, 2008; Cheng and Agterberg, 2009). These anomalies can also be termed singularities, and the geological processes that form them are termed singular processes (Cheng, 1999). These singular processes always result in fractality and multifractality, primarily due to the natural nonlinear attributes involved in these processes (Cheng, 2007; Cheng and Agterberg, 2009). Recent advances in the methods used in fractal and multifractal analysis have included the development of singularity mapping techniques, multifractal interpolation models, and spectrum–area (S–A) fractal models, all of which have been used to separate and identify geochemical anomalies (Cheng, 2007; Zuo and Cheng, 2008; Cheng and Agterberg, 2009; Yuan et al., 2012b). These models can be used not only to describe the fractal and multifractal characteristics of geochemical data but also to identify more effectively strong and weak anomalies from complex geochemical background data.

Here, a study based on multifractal analysis will be presented to show the usefulness of the multifractal analysis for the geochemical exploration data.

4.3.1.1 Geological setting

The NW Junggar area is located in the northwest of Xinjiang Province, China, within the south-western Central Asian Orogenic Belt (Fig. 4.3.1A). The area is divided into the Sawuer–Taerbahatai and Xiemisitaisub regions (Fig. 4.3.1B).

The region is dominated by structures developed during NE–SW and E–W extension, with the latter associated with the development of the Sawuer, Taerbahatai, and Hongguleleng faults in addition to numerous faults and folds in the north of the area. The study area to the south of the Hongguleleng fault also contains ENE–WSW-oriented faults that dip to the NW, and all major faults are associated with well-developed subsidiary fractures, with varying spatial and temporal characteristics

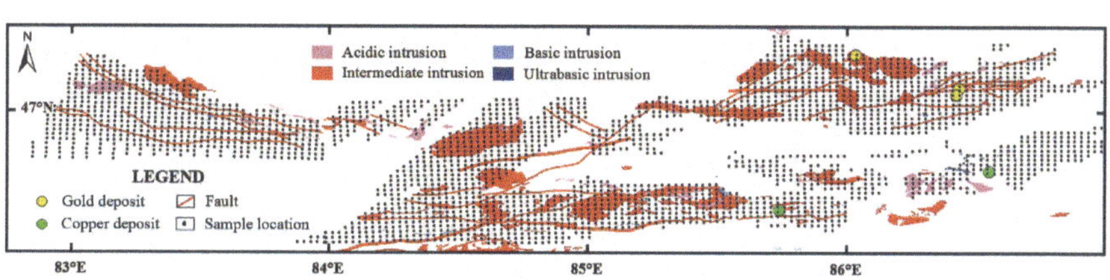

FIGURE 4.3.1 (A) Map showing the location of the study area within the north of Xinjiang Province in an area dominated by the Central Asia Orogenic Belt. *Source:* Modified after Jahn et al. (2000). (B) Simplified geological map of the NW Junggar area, showing major faults and Au and Cu deposits in the study area. *Source: Modified after BGMRXUAR (Bureau of Geology and Mineral Resources of Xinjiang Uygur Autonomous Region), 1993. Regional Geology of Xinjiang Uygur Autonomous Region. Geological Publishing House, Beijing, 841 p. (in Chinese).*

indicative of multiple phases of faulting. These structures play an important role in controlling outcropping lithologies, magmatism, and the type and location of mineralization in the study area (Shen et al., 2007; Chen et al., 2010).

The study area contains Paleozoic pyroclastic, volcaniclastic sedimentary, and volcanic rocks, and widespread intrusive rocks of ultramafic to felsic composition, although the majority of these intrusions are of felsic composition (Zhou et al., 2006; Fan et al., 2007). These intrusions are present as batholiths and stocks, in addition to apophyses and dikes, and the majority of these intrusions are of diorite, quartz diorite, granodiorite, monzogranite, and alkali granite composition (Jahn et al., 2000; Zhou et al., 2006).

Among the discovered deposits in the study area, the Kuoerzhenkuola and Buerkesidai gold deposits are the two largest volcanic-hosted gold deposits. They are hosted in a caldera structure consisting of volcanic and subvolcanic rocks and controlled by a caldera fracture system overprinted by regional faults. After further studies on geology, geochemistry, and geochronology, Shen et al. (2007) proposed that these two volcanic-hosted gold deposits share the same genesis, and both are classified as volcanogenic hydrothermal gold deposits. There have been only a few studies on Xiemisitai and Aermuqiang copper deposits, and Shen et al. (2010) proposed that Xiemisitai copper deposits should be classified as of hydrothermal origin.

4.3.1.2 Method

1. Singularity mapping technique

   Singularity mapping is a technique used to identify geochemical anomalies that result from singular processes (Cheng, 2007; Zuo and Cheng, 2008; Zuo et al., 2015). These singularities can be estimated using a sliding window method that uses the following formula to define the relationship between the average element concentration and the size of the window:

$$X = c \cdot \varepsilon^{\alpha - E} \tag{4.3.1}$$

   where X is the average concentration, c is a constant value, α is the singularity, ε is the size of the window, and E is the Euclidean dimension (Agterberg, 2012).

   The method uses the following steps: (1) defining a set of sliding windows $A(r)$ with both variable and identical interval window sizes, $r_{min} = r_1 < r_2 < \cdots < r_n = r_{max}$; this approach generally uses square window shapes; (2) calculating the singularity at all locations within the study area during steps (3) and (4); (3) calculating the average concentration $C[A(r_i)]$ obtained for each size of window; (4) plotting the average concentration $C[A(r_i)]$ against the size of the window r used to obtain that average concentration on a log–log graph, which should theoretically show a linear relationship (Zuo et al., 2013):

$$\log C[A(r_i)] = c + (\alpha - 2)\log(r) \tag{4.3.2}$$

The singularity value α can be estimated using a linear fitting method from the log–log relationship defined using steps (1)–(4). An α value close to 2 indicates that the given location does not have the characteristics of singularity, whereas locations with values <2 are indicative of areas with element concentration enrichment, and locations with values >2 are indicative of areas with depleted element concentration (Cheng, 2007; Zuo et al., 2013).

2. Multifractal kriging interpolation

Cheng (1999) proposed a multifractal interpolation method using singularity and moving average interpolation to simultaneously measure local singularities and the spatial correlation between data, while also overcoming the smoothing effect of applying a moving average interpolation method. The key principle of this method is the addition of a singularity value to the moving average interpolation method as follows:

$$Z(x_0) = \varepsilon^{2-\alpha(x_0)} \sum_{\Omega(x_0,\varepsilon)} \omega(\|x - x_0\|)Z(x) \tag{4.3.3}$$

where $\alpha(x_0)$ is a singularity at location x_0, and ω is the weight of the moving average function. For a 2D dataset, $\alpha(x_0) = 2$, meaning that the multifractal interpolation method outlined earlier will give a value equal to the moving average interpolation method. In comparison, if x_0 is located in an area of greater than average element concentrations (i.e., areas of enrichment), which contain local singularity characteristics and have $\alpha(x_0)$ values of <2, the multifractal interpolation method will give a larger value than the moving average interpolation method, and if x_0 is located in an area of lower than average element concentrations (i.e., areas of depletion) with $\alpha(x_0)$ values of >2, the multifractal interpolation method above will yield values lower than those obtained using the moving average interpolation method.

3. Spectrum–area fractal modeling

Spectrum–area (S–A) fractal modeling is based on generalized self-similarity theory (Cheng, 2006a), which determines the relationship between spectral energy density values greater than S (spectral energy density), and the area covered by sets of data that have wave number attributes greater than S [$A(>S)$] within a 2D frequency domain, using the following formula (Cheng, 2000; Cheng, 2001, 2006b):

$$A(\geq S) \propto S^{-2d/\beta} \tag{4.3.4}$$

where β is the anisotropic scaling exponent, d is a parameter representing the degree of overall concentration, and $\propto$ denotes proportionality (Zuo et al., 2013). Rearranging this formula for use with 2D data gives the following:

$$A(\geq S) \propto S^{-2d}$$

The differing self-similarity characteristics within the frequency domain mean that two different filters can be used to separate background and anomalous concentrations, here termed the background and anomaly filters.

4.3.1.3 Data

A total of 2854 topsoil samples were collected over a 2 by 2 km grid within the study area, out of which three to five samples were composited into one sample. These soil samples overlie Paleozoic units and intrusions, and the locations of these samples are shown in Fig. 4.3.2.

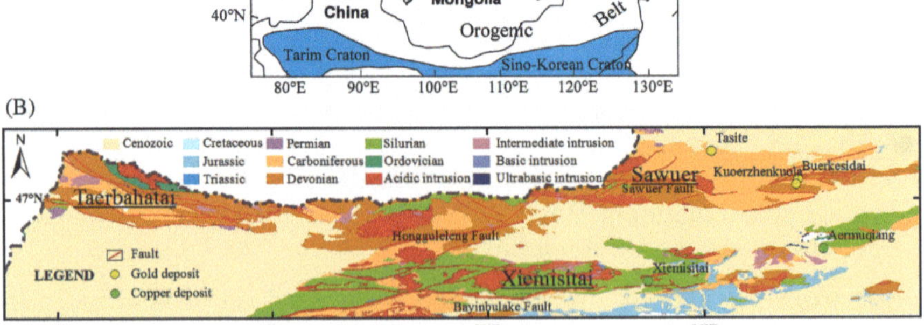

FIGURE 4.3.2 Location map of soil samples, NW Junggar area of northern Xinjiang Province, China.

The concentrations of a total of 39 major and trace elements were measured during this study. Here, the concentrations of As, Cu, and Au within these soil samples are considered, and they are used in the identification of anomalies associated with known Cu and Au mineralization. Arsenic was determined by hydride generation-atomic fluorescence spectrometry, Au by graphite furnace atomic absorption spectrophotometry, and Cu by inductively coupled plasma mass spectrometry.

4.3.1.4 Identification of geochemical anomalies

First, the singularities of the concentrations of As, Cu, and Au within the soil geochemical data set were identified. Based on the singularities, multifractal kriging interpolation method was used to interpolate the

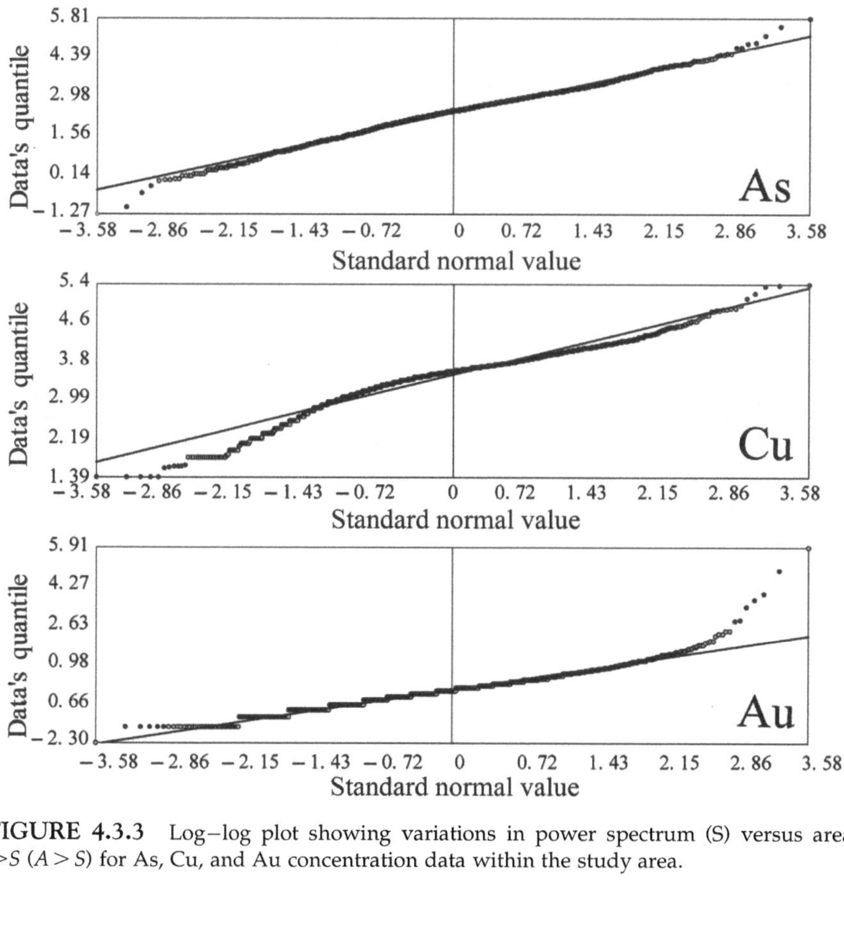

FIGURE 4.3.3 Log–log plot showing variations in power spectrum (S) versus areas >S (A > S) for As, Cu, and Au concentration data within the study area.

soil geochemical data to enhance localized and weak geochemical anomalies. Then, the raster maps produced by interpolation using multifractal kriging were transformed to frequency domain data using a Fourier transformation. Based on an S–A fractal model, the fractal characteristics that represent the relationship between the spectral energy density *S* and the areas >*S* were then analyzed on a log–log plot (Fig. 4.3.3). The As, Cu, and Au data points that are shown in Fig. 4.3.3 can all be modeled using two straight lines. The cut-off value between these two straight lines was identified and used to determine anomaly filters for each element. The final stage was to use the anomaly filter to separate anomalies and areas with background concentrations before translating these anomalies into spatial domain data.

The geochemical anomalies identified by the S–A model are shown in Fig. 4.3.4. All known mineral deposits in this area are associated with distinct geochemical anomalies, and subsequent ground-truthing of anomalies, associated with no known mineralization, directly led to the identification of a number of previously unknown areas of Cu mineralization (Fig. 4.3.5).

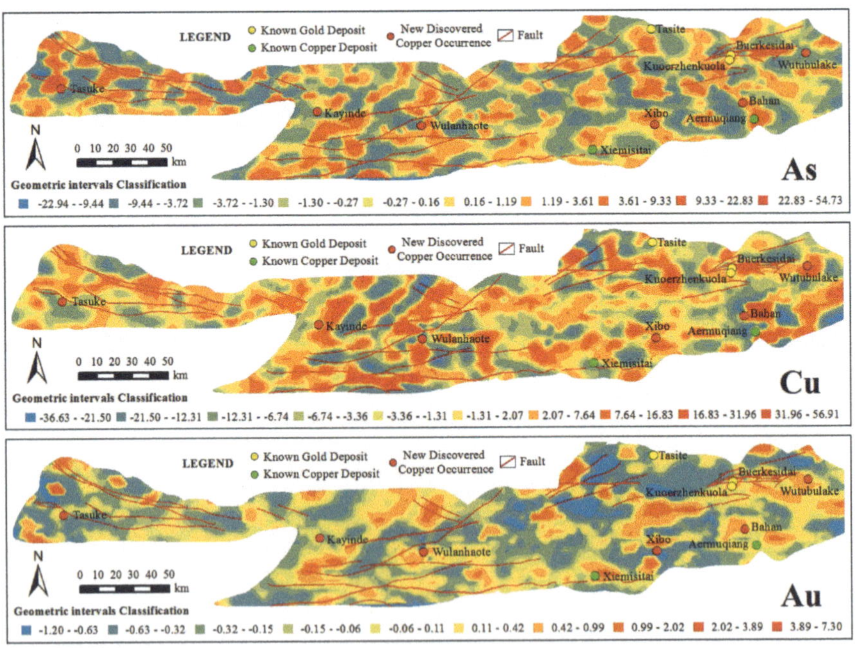

FIGURE 4.3.4 Maps showing the location of As, Au, and Cu geochemical anomalies in the study area determined by S–A fractal modeling compared with the location of known Au and Cu deposits and recently identified Cu prospects, NW Junggar area of northern Xinjiang Province, China.

FIGURE 4.3.5 Photographs from the Bahan copper occurrence. (A) Malachite mineralization within volcanic breccias. (B) Malachite mineralization from the Bahan copper occurrence.

This suggests that the use of fractal and multifractal modeling of soil geochemical for mineral exploration proved to be very effective in the study area in the delineation potential Cu and Au mineralization, and the methodology used is considered to be applicable to other areas around the globe.

Acknowledgments

The researches introduced in this chapter come from the published papers as follows:

Yuan, F., Li, X., Zhou, T., Deng, Y., Zhang D., Xu, C., Zhang R., Jia, C., Jowitt, S.M., 2015. Multifractal modeling-based mapping and identification of geochemical anomalies associated with Cu and Au mineralization in the NW Junggar area of northern Xinjiang Province, China. J. Geochem. Explor. 154, 252–264.

The researches were financially supported by funds from the National Natural Science Foundation of China (Grant No. 41672069), the Chinese National Science and Technology Programme during the 12th Five-year Plan Period (Grant No. 2011BAB06B01), the Program for New Century Excellent Talents in University (Grant No. NCET-10-0324), and the Featured Program for University (Grant No. MS2015HFGY026).

References

326 Geological Team, 2012. Bureau of Geology and Mineral Exploration of Anhui Province. Geological report of Zhuchong Fe-Cu deposit. 243 p. (in Chinese).

Agterberg, F.P., 2012. Multifractals and geostatistics. J. Geochem. Explor. 122, 113–122.

BGMRXUAR (Bureau of Geology and Mineral Resources of Xinjiang Uygur Autonomous Region), 1993. Regional Geology of Xinjiang Uygur Autonomous Region. Geological Publishing House, Beijing, 841 p. (in Chinese).

Bonham-Carter, G., 1994. Geographic Information Systems for Geoscientists: Modelling with GIS. Elsevier, Oxford, 398 p.

Calcagno, P., Chilès, J.P., Courrioux, G., Guillen, A., 2008. Geological modelling from field data and geological knowledge: part I. Modelling method coupling 3D potential field interpolation and geological rules. Phys. Earth Planet. Inter. 171, 147–157.

Chang, Y., Liu, X., Wu, Y., 1991. The Cu, Fe Metallogenic Belt in the Middle–Lower Reaches of Yangtze River. Geological Publishing House, Beijing, 379 p. (in Chinese).

Carranza, E.J.M., 2009. Geochemical anomaly and mineral prospectivity mapping in GIS. In: Hale, M. (Ed.), Handbook of Exploration and Environmental Geochemistry, vol. 11. Elsevier Science Ltd, Oxford, 368 p.

Chen, J., Han, B., Ji, J., Zhang, L., Zhao, X., He, G., et al., 2010. Zircon U–Pb ages and tectonic implications of Paleozoic plutons in northern West Junggar, North Xinjiang, China. Lithos 115, 137–152 (in Chinese with English abstract).

Cheng, Q., 1999. Multifractality and spatial statistics. Comput. Geosci. 25, 949–961.

Cheng, Q., 2000. Multifractal theory and geochemical element distribution pattern. Earth Sci. J. China Univ. Geosci. 25, 311–318 (in Chinese with English abstract).

Cheng, Q., 2001. Spatial self-similarity and geophysical and geochemical anomaly decomposition. Prog. Geophys 16, 8–17.

Cheng, Q., 2006a. Singularity-generalized self-similarity-fractal spectrum (3S) models. Earth Sci. J. China Univ. Geosci. 31, 42–53 (in Chinese with English abstract).

Cheng, Q., 2006b. Multifractal modelling and spectrum analysis: methods and applications to gamma ray spectrometer data from southwestern Nova Scotia, Canada. Sci. China 49, 283–294.

Cheng, Q., 2007. Mapping singularities with stream sediment geochemical data for prediction of undiscovered mineral deposits in Gejiu, Yunnan Province, China. Ore Geol. Rev. 32, 314–324.

Cheng, Q., 2008. Non-linear theory and power-law models for information integration and mineral resources quantitative assessments, progress in geomathematics. In: Bonham-Carter, G.F., Cheng, Q. (Eds.), Progress in Geomathematics. Springer-Verlag, Berlin, Heidelberg, pp. 195–225.

Cheng, Q., Agterberg, F.P., 2009. Singularity analysis of ore-mineral and toxic trace elements in stream sediments. Comput. Geosci. 35, 234–244.

Chung, C., Agterberg, F., 1980. Regression models for estimating mineral resources from geological map data. Math. Geol. 12, 473–488.

Dong, S., Qiu, R., 1993. Tectonism and Magmatism in the Anqing-Yueshan Area. Geological Publish House, Beijing, 158 p (in Chinese).

East China Metallurgical Bureau of Geology and Exploration, 1982. Geological report on the Baixiangshan deposit (in Chinese).

Fabbri, A.G., Chang-Jo, C., 2008. On blind tests and spatial prediction models. Nat. Resour. Res. 17, 107–118.

Fan, Y., Zhou, T., Yuan, F., Tan, L., Cooke, D., Meffre, S., et al., 2007. LA-ICP MS zircon age of Tasite pluton in Sawuer region of west Junggar, Xinjiang. Acta Petrol. Sin. 23, 1901–1908 (in Chinese with English abstract).

Fan, Y., Zhou, T., Yuan, F., Zhang, L., Qian, B., Ma, L., et al., 2011. Geochronology of the porphyry-like type iron deposits in Ning-Wu basin: evidence from $^{40}Ar-^{39}Ar$ phlogopite dating. Acta Petrol. Sin. 85, 810–820 (in Chinese with English abstract).

Itasca Consulting Group Inc. 2012. FLAC3D User's Guide, Version 5.0, Minneapolis.

Jahn, B.-M., Wu, F., Chen, B., 2000. Granitoids of the Central Asian Orogenic Belt and continental growth in the Phanerozoic. Trans. R. Soc. Edinb. Earth. Sci. 91, 181–194.

Jessell, M., 2001. Three-dimensional geological modelling of potential-field data. Comput. Geosci. 27, 455–465.

Jones, M.W., Bærentzen, J.A., Sramek, M., 2006. 3D distance fields: a survey of techniques and applications. IEEE Trans. Vis. Comput. Graph. 12, 581–599.

Li, Y., Oldenburg, D.W., 1996. 3-D inversion of magnetic data. Geophysics 61, 394–408.

Li, Y., Oldenburg, D.W., 1998. 3-D inversion of gravity data. Geophysics 63, 109–119.

Li, X., Yuan, F., Zhang, M., Jia, C., Jowitt, S.M., Ord, A., et al., 2015. Three-dimensional mineral prospectivity modeling for targeting of concealed mineralization within the Zhong gu iron or field, Ning wu Basin, China. Ore Geol. Rev. 71, 633–654.

Li, N., Bagas, L., Li, X., Xiao, K., Li, Y., Ying, L., et al., 2016. An improved buffer analysis technique for model-based 3D mineral potential mapping and its application. Ore Geol. Rev. 76, 94–107.

Li, X., Yuan, F., Zhang, M., Jowitt, S.M., Ord, A., Zhou, T., et al., 2019. 3D computational simulation-based mineral prospectivity modeling for exploration for concealed Fe–Cu skarn-type mineralization within the Yueshan orefield, Anqing district, Anhui Province, China. Ore Geol. Rev. 105, 1–17.

Lin, H.W., Wang, G.J., 2003. Three dimensional signed Euclidean distance transform and its applications. Chin. J. Comp. 26, 1645–1651 (in Chinese with English abstract).

Liu, L., Wan, C., Zhao, C., Zhao, Y., 2011. Geodynamic constraints on orebody localization in the Anqing orefield, China: computational modeling and facilitating predictive exploration of deep deposits. Ore Geol. Rev. 43, 249–263.

Liu, L., Zhao, Y., Sun, T., 2012. 3D computational shape- and cooling process-modeling of magmatic intrusion and its implication for genesis and exploration of intrusion related ore deposits: an example from the Yueshan intrusion in Anqing. Tectonophysics 526, 110–123.

Lü, Q., Qi, G., Yan, J., 2012. 3D geologic model of Shizishan ore field constrained by gravity and magnetic interactive modeling: a case history. Geophysics 78, 25–35.

Malehmir, A., Thunehed, H., Tryggvason, A., 2009. The Paleoproterozoic Kristineberg mining area, northern Sweden: results from integrated 3D geophysical and geologic modeling, and implications for targeting ore deposits. Geophysics 74, 9–22.

Mao, J., Wang, Y., Lehmann, B., Yu, J., Du, A., Mei, Y., et al., 2006. Molybdenite Re-Os and albite 40Ar/39Ar dating of Cu-Au-Mo and magnetite porphyry systems in the Yangtze River valley and metallogenic implications. Ore Geol. Rev. 29, 307–324.

Mao, J., Xie, G., Duan, C., Pirajno, F., Ishiyama, D., Chen, Y., 2011. A tectono-genetic model For Porphyry–Skarn–Stratabound Cu–Au–Mo–Fe and magnetite-apatite deposits along the middle-lower Yangtze River Valley, Eastern China. Ore Geol. Rev. 43, 294–314.

Murphy, F., Ord, A., Hobbs, B., Willetts, G., Barnicoat, A., 2008. Targeting stratiform Zn-Pb-Ag massive sulfide deposits in Ireland through numerical modeling of coupled deformation, thermal transport, and fluid flow. Econ. Geol. 103, 1437–1458.

Ord, A., Hobbs, B.E., Lester, D.R., 2012. The mechanics of hydrothermal systems: I. Ore systems as chemical reactors. Ore Geol. Rev. 49, 1–44.

Pan, G.C., Harris, D.P., 2000. Information Synthesis for Mineral Exploration. Oxford University Press, New York.

Payne, C., Cunningham, F., Peters, K.J., Nielsen, S., Puccioni, E., Wildman, C., et al., 2015. From 2D to 3D: prospectivity modelling in the Taupo volcanic zone, New Zealand. Ore Geol. Rev. 71, 558–577.

Porwal, A., Gonzalez-Alvarez, I., Markwitz, V., McCuaig, T., Mamuse, A., 2010. Weights-of-evidence and logistic regression modelling of magmatic nickel sulfide prospectivity in the Yilgarn Craton, Western Australia. Ore Geol. Rev. 38, 184–196.

Shen, P., Shen, Y., Liu, T., Li, G., Zeng, Q., 2007. Genesis of volcanic-hosted gold deposits in the Sawur gold belt, northern Xinjiang, China: evidence from REE, stable isotopes, and noble gas isotopes. Ore Geol. Rev. 32, 207–226.

Shen, P., Shen, Y., Tie-bing, L., Pan, H.-D., Meng, L., Song, G., Dai, H., 2010. Discovery of the xiemisitai copper deposit in western Junggar, xinjiang and its geological significance. Xinjiang Geol. 28, 413–418 (In Chinese with English abstract).

Sun, W., Yuan, F., Jowitt, S.M., Zhou, T., Hollings, P., Liu, G., et al., 2017. Geochronology and geochemistry of the Fe ore-bearing Zhonggu intrusions of the Ningwu Basin: implications for tectonic setting and contemporaneous Cu-Au mineralization in the Middle-Lower Yangtze Metallogenic Belt. Ore Geol. Rev. 84, 246–272.

Wang, W., Zhao, J., Cheng, Q., 2011. Analysis and integration of geo-information to identify granitic intrusions as exploration targets in southeastern Yunnan District, China. Comp. Geosci. 37, 1946–1957.

Williams, N.C., 2008. Geologically-constrained UBC–GIF gravity and magnetic inversions with examples from the Agnew-Wiluna greenstone belt, Western Australia. Ph.D. Dissertation. The University of British Columbia, Kelowna, 488 p.

Xing, F., Xu, X., Li, Z., 1994. Discovery of the Early Proterozoic basement in the Middle-Lower Reaches of Yangtze River and its significance. Chin. Sci, Bull. 39, 135–139.

Yuan, F., Zhou, T., Wang, S., Fan, Y., Tang, C., Zhang, Q., et al., 2012a. Characteristics of alteration and mineralization of the Shaxi porphyry copper deposit, Luzong area, Anhui Province. Acta Petrol. Sin. 28, 3099–3112 (in Chinese with English abstract).

Yuan, F., Li, X., Jowitt, S.M., Zhang, M., Jia, C., Bai, X., et al., 2012b. Anomaly identification in soil geochemistry using multifractal interpolation: a case study using the distribution of Cu and Au in soils from the Tongling mining district, Yangtze Metallogenic Belt, Anhui Province, China. J. Geochem. Explor. 116–117, 28–39.

Yuan, F., Li, X., Zhang, M., Jowitt, S.M., Jia, C., Zheng, T., et al., 2014. Three-dimensional weights of evidence-based prospectivity modeling: a case study of the Baixiangshan mining area, Ningwu Basin, Middle and Lower Yangtze Metallogenic Belt, China. J. Geochem. Explor. 145, 82–97.

Yuan, F., Li, X., Zhou, T., Deng, Y., Zhang, D., Xu, C., et al., 2015. Multifractal modeling-based mapping and identification of geochemical anomalies associated with Cu and Au mineralization in the NW Junggar area of northern Xinjiang Province, China. J. Geochem. Explor. 154, 252–264.

Zhai, Y., Yao, S., Lin, X., Zhou, X., Wan, T., Jin, F., et al., 1992. The Metallogeny of the Fe–Cu (Au) Deposits in the Middle–Lower Yangtze Region. Geological Publishing House, Beijing, 235 p. (in Chinese).

Zhou, T., Yuan, F., Tan, L., Fan, Y., Yue, S., 2006. Geodynamic significance of the A-type granites in the Sawuer region in west Junggar, Xinjiang: rock geochemistry and SHRIMP zircon age evidence. Sci. China Ser. D 29, 113–123.

Zhou, T., Yuan, F., Yue, S., Liu, X., Zhang, X., Fan, Y., 2007. Geochemistry and evolution of ore-forming fluids of the Yueshan Cu–Au skarn- and vein-type deposits, Anhui Province, South China. Ore Geol. Rev. 31, 279–303.

Zhou, T., Fan, Y., Yuan, F., Zhang, L., Qian, B., Ma, L., et al., 2011. Geochronology and significance of volcanic rocks in the Ning-Wu Basin of China. Sci. China Earth Sci. 41, 960–971.

Zhou, T., Wang, S., Fan, Y., Yuan, F., Zhang, D., White, N.C., 2015. A review of the intracontinental porphyry deposits in the Middle-Lower Yangtze River Valley Metallogenic Belt, Eastern China. Ore Geol. Rev. 65, 433–456.

Zuo, R., Cheng, Q., 2008. Mapping singularities—a technique to identify potential Cu mineral deposits using sediment geochemical data, an example for Tibet, west China. Miner. Mag. 72, 531–534.

Zuo, R., Xia, Q., Zhang, D., 2013. A comparison study of the C–A and S–A models with singularity analysis to identify geochemical anomalies in covered area. Appl. Geochem. 33, 165–172.

Zuo, R., Wang, J., Chen, G., Yang, M., 2015. Identification of weak anomalies: a multifractal perspective. J. Geochem. Explor. 148, 12–24.

Index

Note: Page numbers followed by "*f*" and "*t*" refer to figures and tables, respectively.

Y

Z

FSC
www.fsc.org
MIX
Papier | Fördert
gute Waldnutzung
FSC® C083411

Kolibri 360 GmbH
Ferdinand-Jühlke-Straße 7
99095 Erfurt, Deutschland
produktsicherheit@kolibri360.de